A LEVEL
Questions and Answers

MATHEMATICS

Peter Sherran & Janet Crawshaw

SERIES EDITOR: BOB McDUELL

Contents

HOW TO USE THIS BOOK

The aim of the *Questions and Answers* series is to provide students with the help required to attain the highest level of achievement in important examinations. This book is intended to help you with A- and AS-level Mathematics or, in Scotland, Higher Level Mathematics. The series relies on the idea that an experienced examiner can provide, through examination questions, sample answers and advice, the help students need to secure success. Many revision aids concentrate on providing factual information that might have to be recalled in an examination. This series, while giving factual information in an easy-to-remember form, concentrates on the other skills that need to be developed for the new A-level examinations being introduced from 1996.

The *Questions and Answers* series is designed to provide:

- Easy-to-use **Revision Summaries** that identify important factual information that students must understand if progress is to be made in answering examination questions.

- Advice on the different types of question in each subject and how to answer them well to obtain the highest marks.

- Information about other skills, apart from the recall of knowledge, that will be tested on examination papers. These are sometimes called **assessment objectives** and modern A-level examinations put great emphasis on them. The *Questions and Answers* series is intended to develop these skills, particularly of communication, problem-solving, evaluation and interpretation, by the use of questions and the appreciation of outcomes by the student.

- Many examples of **examination questions**. Students can increase their achievement by studying a sufficiently wide range of questions, provided that they are shown the way to improve their answers to these questions. It is advisable that students try the questions first before looking at the answers and the advice that accompanies them. All the Mathematics questions come from actual examination papers or specimen materials issued by the British Examination Boards, reflecting their requirements.

- **Sample answers** and mark schemes to all the questions.

- **Advice from Examiners**: by using the experience of actual examiners we are able to give advice that can enable students to see how their answers can be improved to ensure greater success.

Success in A-level examinations comes from proper preparation and a positive attitude, developed through a sound knowledge of facts and an understanding of principles. These books are intended to overcome 'examination nerves' which often come from a fear of not feeling properly prepared.

THE IMPORTANCE OF USING QUESTIONS FOR REVISION

Past examination questions play an important part in revising for examinations. However, it is important not to start practising questions too early. Nothing can be more disheartening than trying to do a question that you do not understand because you have not mastered the concepts. Therefore it is important to have studied a topic thoroughly before attempting questions on it.

It is unlikely that any question you try will appear in exactly the same form on the papers you are going to take. However the number of totally original questions that can be set on any part of the syllabus is limited and so similar ideas occur over and over again. It certainly will help you if the question you are trying to answer in an examination is familiar and you are used to the type of language used. Your confidence will be boosted, and confidence is important for examination success.

Practising examination questions will also highlight gaps in your knowledge and understanding that you can go back and revise more thoroughly. It will indicate which sorts of question you can do well and which, if there is a choice, you should avoid.

Finally, having access to answers, as you do in this book, will enable you to see clearly what is required by the examiner, how best to answer each question and the amount of detail required. Remember that attention to detail is a very important aspect of achieving success at A-level.

MAXIMISING YOUR MARKS IN MATHEMATICS

One of the keys to examination success is to know how marks are gained or lost and the examiner's tips given with the solutions in this book give hints on how you can maximise your marks on particular questions. However you should also take careful note of these general points:

- Check the requirements of your examination board and follow the instructions (or 'rubric') carefully about the number of questions to be tackled. Many A-level Mathematics examinations instruct you to attempt all the questions and where papers start with short, straightforward questions, you are advised to work through them in order so that you build up your confidence. Do not overlook any parts of a question – double-check that you have seen everything, including any questions on the back page! If there is a choice, do the correct number. If you do more, you will not be given credit for any extra and it is likely that you will not have spent the correct time on each question and your answers could have suffered as a result. Take time to read through all the questions carefully, and then start with the question you think you can do best.

- Get into the habit of setting out your work neatly and logically. If you are untidy and disorganised you could penalise yourself by misreading your own figures or lose marks because your method is not obvious. Always show all necessary working so that you can obtain marks for a correct method even if your final answer is wrong. Remember that a good clear sketch can help you to see important details.

- When the question asks for a particular result to be established, remember that to obtain the method marks you must show sufficient working to convince the examiner that your argument is valid. Do not rely too heavily on your graphical calculator.

- Do not be sloppy with algebraic notation or manipulation, especially involving brackets and negatives. Do rough estimates of calculations to make sure that they are reasonable, state units if applicable and give answers to the required degree of accuracy; do not approximate prematurely in your working.

- Make sure that you are familiar with the formulas booklet and tables that you will be given in the examination and learn any useful formulas that are not included. Refer to the booklet in the examination and transfer details accurately.

- When about 15 minutes remain, check whether you are running short of time. If so, try to score as many marks as possible in the short time that remains, concentrating on the easier parts of any questions not yet tackled.

- The following glossary may help you in answering questions:
  Write down, state – no justification is needed for an answer.
  Calculate, find, determine, show, solve – include enough working to make your method clear.
  Deduce, hence – make use of the given statement to establish the required result.
  Sketch – show the general shape of a graph, its relationship with the axes, any asymptotes and points of special significance such as turning points.
  Draw – plot accurately, using graph paper and selecting a suitable scale; this is usually preparation for reading information from the graph.
  Find the <u>exact</u> value – leave it in fractions or surds, or in terms of logarithms, exponentials or π; note that using a calculator is likely to introduce decimal approximations, resulting in loss of marks.

Rules for indices

$$a^m \times a^n = a^{m+n} \qquad a^m \div a^n = a^{m-n} \qquad a^{mn} = (a^m)^n = (a^n)^m \qquad a^{\frac{1}{n}} = \sqrt[n]{a}$$

$$a^{\frac{m}{n}} = (a^m)^{\frac{1}{n}} = (a^{\frac{1}{n}})^m \qquad a^{-n} = \frac{1}{a^n} \qquad (ab)^n = a^n b^n \qquad \left(\frac{a}{b}\right)^n = \frac{a^n}{b^n}$$

Note, however, in general $(a+b)^n \neq a^n + b^n$ and $(a-b)^n \neq a^n - b^n$

Some important special cases are: $a^1 = a$ and, provided $a \neq 0$, $a^0 = 1$.

$$a^{-1} = \frac{1}{a} \quad \text{and} \quad a^{\frac{1}{2}} = \sqrt{a} \quad \text{(Note: } a^{\frac{1}{2}} \times a^{\frac{1}{2}} = a^1 = a \text{)}.$$

Logarithm is another word for **power, exponent** or **index**. The statement $10^2 = 100$ may be described by saying that the logarithm associated with 100, taking 10 as the **base**, is 2. This is usually written as $\log_{10} 100 = 2$. In general, $a^x = y \Leftrightarrow \log_a y = x$.

Natural logarithms have base $e = 2.718\ 28...$ and are needed in order to integrate some types of **rational function**. The notation for natural logarithms is $\log_e x$ or $\ln x$.

The three basic laws of logarithms are: $\qquad \log_a x + \log_a y = \log_a xy$

$$\log_a x - \log_a y = \log_a \frac{x}{y}$$

$$\log_a x^n = n \log_a x$$

Logarithms may be used to solve **exponential equations** i.e. equations in which the unknown value is the exponent and to convert the graphs of exponential functions to **linear form**.

A function which is the sum of terms of the form ax^n, where n is a non-negative whole number and a is a constant, is called a **polynomial**. The highest value of n that occurs in the polynomial is known as the **degree** of the polynomial.

The remainder theorem states that when any polynomial function $f(x)$ is divided by $(x - r)$ the remainder is given by $f(r)$.

This approach enables us to find the remainder much quicker than by carrying out the division.

The factor theorem is simply a special case of this result corresponding to a remainder of zero. For any polynomial function $f(x)$, if $f(r) = 0$ then $(x - r)$ is a **factor** of $f(x)$.

The factor theorem may be applied to the solution of polynomial equations and to the sketching of associated graphs.

The **modulus** of a function $f(x)$ is denoted by $|f(x)|$ and takes the positive numerical value of $f(x)$.

This notation allows a condition such as $-1 < x < 1$ to be expressed simply as $|x| < 1$. The distance between two points, a and b, on the number line is given by the positive difference between a and b which may be written as $|a - b|$.

The graph of $y = |f(x)|$ is the same as the graph of $y = f(x)$ wherever $f(x) \geq 0$, but points where $f(x) < 0$ are reflected in the x-axis.

The general form of a **quadratic** function is $f(x) = ax^2 + bx + c$ (where $a \neq 0$). The graph of a quadratic function is always a **parabola**, which takes one of two possible forms depending on the sign of a (the coefficient of x^2).

Fig. 1 $a > 0$ $a < 0$

The solutions of the quadratic equation $f(x) = 0$ are given by the intersection of the graph with the *x*-axis. These may be located approximately using a graphics calculator or computer graph plotter.

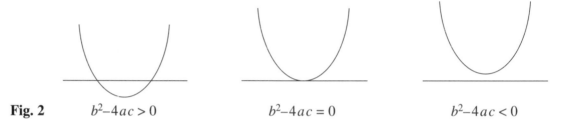

Fig. 2 $b^2 - 4ac > 0$ $b^2 - 4ac = 0$ $b^2 - 4ac < 0$

The diagrams above correspond to the situation where $a > 0$, but the principles apply equally when $a < 0$.

The solution of some quadratic equations may be found by **factorisation**.

$$2x^2 - 5x - 3 = 0 \Rightarrow (2x + 1)(x - 3) = 0 \Rightarrow 2x + 1 = 0 \text{ or } x - 3 = 0 \Rightarrow x = -\tfrac{1}{2} \text{ or } x = 3.$$

An alternative method that may be employed, even when factorisation is not possible, is known as **completing the square**. This involves re-writing an expression of the form $x^2 + bx + c$ as $(x + p)^2 + q$, which has the key advantage that the *unknown value only appears once*.

$$x^2 + 6x - 5 = 0 \Rightarrow (x + 3)^2 - 9 - 5 = 0 \qquad \text{Note: } (x + 3)^2 = x^2 + 6x + 9$$
$$\Rightarrow (x + 3)^2 - 14 = 0$$
$$\Rightarrow x + 3 = \pm\sqrt{14}$$
$$\Rightarrow x = -3 + \sqrt{14} \text{ or } x = -3 - \sqrt{14}$$

Note: $p = \tfrac{1}{2}b$ and $q = c - p^2$. However, if the process is applied to the general

quadratic $ax^2 + bx + c = 0$ then this leads to the formula $x = \dfrac{-b \pm \sqrt{b^2 - 4ac}}{2a}$.

In the formula, the value of $b^2 - 4ac$ is known as the **discriminant** and may be used to determine the nature of the solutions as shown in Fig. 2.

$b^2 - 4ac > 0$ *two* distinct **real** solutions (the solutions are often referred to as **roots**).

$b^2 - 4ac = 0$ *one* real solution (often regarded as *two* **equal** roots)

$b^2 - 4ac < 0$ *no* real solutions (but there are two **complex** roots).

Another application of completing the square is in connection with **maximum** and **minimum** values of quadratic functions.

The *minimum* value of $(x + p)^2 + q$ occurs when $(x + p)^2 = 0$ i.e. when $x = -p$. It follows that q is the minimum value and that the lowest point on the graph of the function has coordinates $(-p, q)$.

Note: A function of the form $q - (x + p)^2$ has *maximum* value q when $x = -p$.

The most common *algebraic* approach to solving **simultaneous equations** initially involves reducing the set of equations to a single equation in one unknown. The two main methods of achieving this are:

❶ The **elimination** method – multiples of the equations are either added or subtracted in order to eliminate one of the unknown values.

$$2x + 3y = 7 \qquad (1)$$
$$3x - y = 5 \qquad (2)$$

Check that (1) + 3 × (2) gives $\qquad 11x = 22 \Rightarrow x = 2.$

Substituting for x in (2) now gives $\quad 6 - y = 5 \Rightarrow y = 1.$
(*Check* that these values work in (1)).

❷ The **substitution** method – one of the unknowns is made the subject of one of the equations and the result is substituted wherever the unknown appears.

From (2) $y = 3x - 5.$ Substituting for y in (1) gives $2x + 3(3x - 5) = 7 \Rightarrow 11x - 15 = 7.$ From this point, the solution proceeds as before. The substitution method can also be used, for example, when only one equation is linear.

The manipulation of **inequalities** is much the same as the manipulation of equations apart from the complication that whenever a negative factor is introduced to both sides, the direction of the inequality is reversed. For example $-x > 2 \Rightarrow x < -2.$ Care must also be taken when carrying out steps such as squaring both sides whenever negatives are involved.

If you need to revise this subject more thoroughly, see the relevant topics in the _Letts_ A level Mathematics Study Guide.

Note: As an alternative to the strictly algebraic approach to the solution of equations and inequalities, graphs can be used to determine the nature, and approximate value, of the solutions. A graphics calculator is particularly useful, in this respect, and its trace and zoom facilities can often be used to locate these solutions to a high degree of accuracy. It is important, however, to include *details of the method* used in any solution.

Some functions of the form $\dfrac{p(x)}{q(x)}$ (where $p(x)$ and $q(x)$ are polynomials) are best expressed in **partial fractions** before we attempt to integrate them or work out their series expansions. The following forms, where the degree of $p(x) <$ degree $q(x)$, are particularly important:

$$\frac{p(x)}{(ax+b)(cx+d)} \equiv \frac{A}{ax+b} + \frac{B}{cx+d} \qquad \frac{p(x)}{(ax+b)(cx+d)^2} \equiv \frac{A}{ax+b} + \frac{B}{cx+d} + \frac{C}{(cx+d)^2}$$

$$\frac{p(x)}{(ax+b)(cx^2+d)} \equiv \frac{A}{ax+b} + \frac{Bx+C}{cx^2+d} \qquad \text{(You can \emph{check} by drawing the graph of each side).}$$

Note: If degree $p(x) \ge$ degree $q(x)$ *then these results do not hold* and division must be carried out first.

1 *Algebra*

QUESTIONS

1 (i) Given that $a^k = \sqrt[3]{(a^4)} \div a$, find the value of k.

 (ii) Given that $27^x = 9^{(x-1)}$, find the value of x. (8)

ULEAC

2 (a) Solve the equation $2x^{\frac{1}{3}} = x^{-\frac{2}{3}}$. (4)

 (b) Use your calculator to find the value of y, correct to three significant figures, where $3^y = 6$. (5)

Oxford & Cambridge (MEI)

3 Two variable quantities x and y are related by the equation $y = a(b^x)$, where a and b are constants. When a graph is plotted showing values of $\ln y$ on the vertical axis and values of x on the horizontal axis, the points lie on a straight line having gradient 1.8 and crossing the vertical axis at the point $(0, 4.1)$. Find the values of a and b. (5)

UCLES

4 The population of a country has been recorded at 10 year intervals during this century. The figures, in millions given to the nearest million, for the years 1910 to 1960, are as follows.

Year	1910	1920	1930	1940	1950	1960
Population	37	44	53	64	77	93

One model for population size is given by $P_n = P_0 \times k^n$

where P_0 is the population at some starting date

 P_n is the population n years later

and k is a constant.

 (i) According to this model, what happens to the population in each of the cases

 (A) $k > 1$ (B) $k = 1$ (C) $k < 1$? (3)

 (ii) Complete the table below giving $\log_{10} P_n$ for different values of n. Notice that P_0 is taken to be the population in the year 1900 and so n is the number of years that have elapsed since 1900. (2)

Year	1910	1920	1930	1940	1950	1960
n	10	20				60
$\log_{10} P_n$	1.57					1.97

 (iii) Draw the graph of $\log_{10} P_n$ against n on the axes given opposite. (2)

 (iv) Explain how you can tell from the graph whether the given model gives a good description of the population. (2)

 (v) Estimate the values of P_0 and k from your graph. (3)

The populations in 1970 to 1990 were as follows.

Year	1970	1980	1990
Population	110	125	138

 (vi) Comment on the significance of these figures, bearing in mind the given model. (2)

6

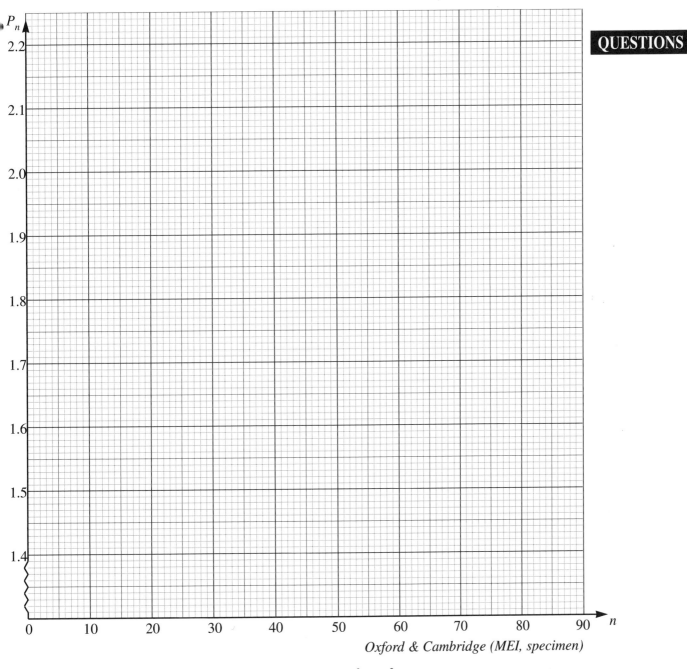

Oxford & Cambridge (MEI, specimen)

5   Given that $(x-2)$ and $(x+2)$ are each factors of $x^3 + ax^2 + bx - 4$, find the values of a
and b.   (4)

  For these values of a and b, find the other linear factor of $x^3 + ax^2 + bx - 4$.   (2)
                                                                                           UCLES

6   (a) Show that $(x+2)$ is a factor of the polynomial $f(x)$ given by

$$f(x) = 2x^3 - 3x^2 - 11x + 6.$$ (1)

  (b) Express $f(x)$ as the product of three linear factors.   (2)

  (c) By considering the graph of $y = f(x)$, or otherwise, solve the inequality $f(x) \le 0$.   (2)
                                                                                            NEAB

7   Express $x^2 - 4x + 9$ in the form $(x-a)^2 + b$, where a and b are constants. Hence, or

  otherwise, state the maximum value of $f(x) = \dfrac{1}{x^2 - 4x + 9}$.   (3)
                                                                                             AEB

1 Algebra

8 Given that $(2x+1)$ is a factor of $2x^3 + ax^2 + 16x + 6$, show that $a = 9$. (2)

Find the real quadratic factor of $2x^3 + 9x^2 + 16x + 6$. By completing the square, or otherwise, show that this quadratic factor is positive for all real values of x. (4)

UCLES

9 (i) Sketch the graph of the function f where $f : x \longrightarrow |3x - 1|$ for all $x \in \mathbb{R}$ (3)

(ii) Find the *two* values of x such that $|3x - 1| = x$. (3)

(iii) Hence find the range of values of x for which $|3x - 1| > x$. (3)

NICCEA

10 (a) On the same diagram, sketch the graphs of

$$y = \frac{1}{x-a} \quad \text{and} \quad y = 4|x-a|, \text{ where } a \text{ is a positive constant.}$$

Show clearly the coordinates of any points of intersection with the coordinate axes.

(b) Hence or otherwise, find the set of values of x for which $\dfrac{1}{x-a} < 4|x-a|$. (12)

ULEAC

11 Find the values of k for which the equation $2x^2 + 4x + k = 0$ has real roots. (2)

SEB

12 The quadratic equation $x^2 + 6x + 1 = k(x^2 + 1)$ has equal roots.
Find the possible values of the constant k. (4)

AEB

13 (a) The point $A(2, 2)$ lies on the parabola $y = x^2 + px + q$.

Find a relationship between p and q. (1)

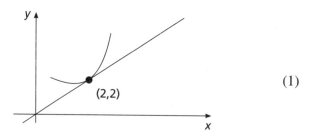

(b) The tangent to the parabola at A is the line $y = x$. Find the value of p.

Hence find the equation of the parabola. (6)

(c) Using your answers for p and q, find the value of the discriminant of $x^2 + px + q = 0$.

What feature of the above sketch is confirmed by this value? (2)

SEB

14 Solve the simultaneous equations

$$x + y = 2$$
$$x^2 + 2y^2 = 11.$$
(6)

UCLES

15 The straight line $y = x$ cuts the circle $x^2 + y^2 - 6x - 2y - 24 = 0$ at A and B.

(a) Find the coordinates of A and B. (3)

(b) Find the equation of the circle which has AB as diameter. (3)

SEB

16 Solve the pair of simultaneous equations $\log_e(x + y) = 0$

$$2\log_e x = \log_e(y - 1).$$ (10)

NICCEA

17 (a) Express $\dfrac{1 - x - x^2}{(1 - 2x)(1 - x)^2}$ as the sum of three partial fractions. (4)

(b) Hence or otherwise, expand this expression in ascending powers of x up to and including the term in x^3. (5)

(c) State the range of values of x for which this expansion is valid. (1)

NEAB

18 (a) You are given that $f(x) = 2x^3 - x^2 - 7x + 6$.

(i) Show that $f(1) = 0$.

Hence find the three factors of $f(x)$. (4)

(ii) Solve the inequality $f(x) > 0$. (3)

(b) (i) Given that

$$\frac{x^2 + 2x + 7}{(2x + 3)(x^2 + 4)} \equiv \frac{A}{(2x + 3)} + \frac{Bx + C}{(x^2 + 4)}$$

find the values of the constants A, B and C. (3)

(ii) Use your answer to (b) (i) to find

$$\int \frac{x^2 + 2x + 7}{(2x + 3)(x^2 + 4)} \, dx.$$ (3)

Oxford & Cambridge (MEI)

19 $f(x) \equiv \dfrac{x^2 + 6x + 7}{(x + 2)(x + 3)}, \ x \in \mathbb{R}$

Given that $f(x) \equiv A + \dfrac{B}{(x + 2)} + \dfrac{C}{(x + 3)}$,

(a) find the values of the constants A, B and C,

(b) show that $\displaystyle\int_0^2 f(x)\,dx = 2 + \ln\left(\frac{25}{18}\right)$. (11)

ULEAC

2 *Sequences and series*

REVISION SUMMARY

A list of numbers in a particular order, and subject to some rule for obtaining subsequent values, is called a **sequence**. Each number in a sequence is called a **term**, and terms are often denoted by $u_1, u_2, u_3, ..., u_n, ...$

A sequence may be defined by

❶ Using a formula for the **general term** .

For example, substituting $n = 1, 2, 3, ...$ into the formula $u_n = n^2 + 1$ generates the sequence 2, 5, 10, 17, 26, ... In this way, the value of any term may be calculated directly by substituting its position into the formula. The 100th term, for example, would be given by $100^2 + 1 = 10\,001$.

❷ Using an **inductive definition**.

This describes how a given term relates to the previous term or terms e.g $u_{n+1} = 2u_n$. Such a description is known as a **recurrence relation** and requires knowledge of the value of some term, or terms, to be put into effect. In this example, given that $u_1 = 3$ would generate the sequence 3, 6, 12, 24, ... The **Fibonacci sequence** 1, 1, 2, 3, 5, 8, ..., on the other hand, may be defined by $u_{n+2} = u_{n+1} + u_n$ where $u_1 = 1, u_2 = 1$ i.e. knowledge of two consecutive terms is needed.

Sequences which carry on indefinitely are either convergent or divergent. In **convergent** sequences, terms approach a particular value known as a **limit**. The following sequences are convergent.

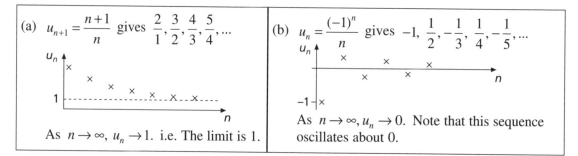

(a) $u_{n+1} = \dfrac{n+1}{n}$ gives $\dfrac{2}{1}, \dfrac{3}{2}, \dfrac{4}{3}, \dfrac{5}{4}, ...$

As $n \to \infty, u_n \to 1$. i.e. The limit is 1.

(b) $u_n = \dfrac{(-1)^n}{n}$ gives $-1, \dfrac{1}{2}, -\dfrac{1}{3}, \dfrac{1}{4}, -\dfrac{1}{5}, ...$

As $n \to \infty, u_n \to 0$. Note that this sequence oscillates about 0.

A sequence that does not converge to a limit is said to be **divergent**. The following sequences are divergent.

(a) $u_{n+1} = 2u_n, u_1 = 1$
gives 1, 2, 4, 8, 16, ...

This sequence may also be defined by $u_n = 2^{n-1}$.

(b) $u_n = \sin\left(\dfrac{n\pi}{2}\right)$
gives 1, 0, –1, 0, 1, 0, –1, 0, ...

After every 4 terms, this sequence repeats itself. Its behaviour is described as **periodic** (with **period** 4).

(c) $u_{n+1} = 5 - u_n, u_1 = 2$
gives 2, 3, 2, 3, 2, 3, ...

Note that this sequence is both **oscillatory** (about 2.5) and periodic (with period 2).

Two special sequences are the **arithmetic progression** (A.P.) and the **geometric progression** (G.P.) In an A.P. successive terms have a **common difference** e.g. 1, 4, 7, 10, ...
In the usual notation, the first term and the common difference are denoted by a and d respectively. The inductive definition of an A.P. could be given as $u_{n+1} = u_n + d, u_1 = a$.

Thus, an A.P. takes the form $a, a + d, a + 2d, a + 3d, ...$ and the nth term is $u_n = a + (n-1)d$.

Letts
Q&A

REVISION
SUMMARY

The inductive definition of a G.P. could be given as $u_{n+1} = ru_n$, $u_1 = a$. Thus a G.P. takes the form a, ar, ar^2, ar^3, ar^4, ... where r is called the **common ratio** e.g. 1, 3, 9, 27, ... The nth term of a G.P. is given by $u_n = ar^{n-1}$.

A **series** is formed by adding together the terms of a sequence. The use of **sigma notation** can greatly simplify the way that series are written. For example, the series $1^2 + 2^2 + 3^2 + 4^2 + ... + n^2$

may be written as $\sum_{i=1}^{n} i^2$. The sum of the first n terms of a series is often denoted by S_n and so

$$S_n = u_1 + u_2 + u_3 + ... + u_n = \sum_{i=1}^{n} u_i.$$

The sum of an A.P. is given by $S_n = \dfrac{n}{2}(2a + (n-1)d)$ which may be written as $S_n = \dfrac{n}{2}(a+l)$ where l is the last term. In a given situation, one form may be more convenient to use than the other depending on the information available.

The sum of a G.P. is given by $S_n = \dfrac{a(1-r^n)}{1-r}$ or alternatively by $S_n = \dfrac{a(r^n-1)}{r-1}$. The choice of which formula to use depends on the value of r and only amounts to avoiding minus signs. Provided that $|r| < 1$, the sum of a G.P. converges to $\dfrac{a}{1-r}$ as $n \to \infty$. This is sometimes written as $S_\infty = \dfrac{a}{1-r}$.

When n is a positive integer, the **binomial expansion** of $(1+x)^n$ is given by

$$(1+x)^n = 1 + nx + \frac{n(n-1)x^2}{2!} + \frac{n(n-1)(n-2)x^3}{3!} + ...$$

In this case, the expansion contains $n+1$ terms (the last of which is x^n) and is valid for *all x.*

Each term is of the form $\dbinom{n}{r} x^r$ where $\dbinom{n}{r} = \dfrac{n(n-1)...(n-r+1)}{r!}$ and so $(1+x)^n = \sum_{r=0}^{n} \dbinom{n}{r} x^r$.

The first term corresponds to $r = 0$, the second term to $r = 1$ and so on. (Note that 0! = 1). This form tends to be used more for finding particular terms than for working out the full expansion.

If n is *not* a positive integer then the expansion continues indefinitely and is *only valid for $|x| < 1$.*

e.g. Expanding $(1-x)^{-1}$ we obtain the series $1 + x + x^2 + x^3 + x^4 + ...$

The significance of the condition that $|x| < 1$ might be seen by substituting particular values of x.

When $x = \frac{1}{2}$ this produces $2 = 1 + \frac{1}{2} + \frac{1}{4} + \frac{1}{8} + ...$ (Note: this is the result we would expect for the sum of an infinite G.P. with $a = 1$ and $r = \frac{1}{2}$).

However, taking $x = 2$, $(1-x)^{-1} = -1$ and this is *not* the same as

$1 + x + x^2 + x^3 + ... = 1 + 2 + 2^2 + 2^3 + ...$ which is seen to be divergent.

For *small values of x*, the first few terms of an infinite series may provide a good

approximation to a function. For example, using the binomial expansion, $\sqrt{1+x} \approx 1 + \dfrac{x}{2} - \dfrac{x^2}{8}$.

Some important examples, based on the **Maclaurin expansion**, are:

$\sin x \approx x$, $\cos x \approx 1 - \dfrac{x^2}{2}$, $\tan x \approx x$. Note that, in each case, x is in **radians**.

If you need to revise this subject more thoroughly, see the relevant topics in the *Letts* A level *Mathematics Study Guide.*

2 Sequences and series

1 (i) The tenth term of an arithmetic progression is 36, and the sum of the first ten terms is 180. Find the first term and the common difference. (4)

(ii) Evaluate $\displaystyle\sum_{r=1}^{1000}(3r-1)$. (3)

UCLES

2 (a) A geometric progression has non-zero first term a and common ratio r, where $0 < r < 1$. Given that the sum of the first 8 terms of the progression is equal to half the sum to infinity, find the value of r, correct to 3 decimal places. (3)

Given also that the 17th term of the progression is 10, find a. (2)

(b) An arithmetic progression has first term a and common difference 10. The sum of the first n terms of the progression is 10 000. Express a in terms of n, and show that the nth term of the progression is

$$\frac{10\ 000}{n}+5(n-1).$$ (3)

Given that the nth term is less than 500, show that $n^2-101n+2000<0$, and hence find the largest possible value of n. (4)

UCLES

3 A small ball is dropped from a height of 1 m on to a horizontal floor. Each time the ball strikes the floor it rebounds to $\frac{3}{5}$ of the height it has fallen.

(a) Show that, when the ball strikes the floor for the third time, it has travelled a distance of 2.92 m.

(b) Show that the total distance travelled by the ball cannot exceed 4 m. (7)

ULEAC

4 Given that $(1+kx)^8 = 1+12x+px^2+qx^3+...$, for all $x \in \mathbb{R}$

(a) find the value of k, the value of p and the value of q.

(b) Using your values of k, p and q find the numerical coefficient of the x^3 term in the expansion of $(1-x)(1+kx)^8$. (11)

ULEAC

5 Write down the expansion of $(1+x)^5$. (1)

Hence, by letting $x = z + z^2$, find the coefficient of z^3 in the expansion of $(1+z+z^2)^5$ in powers of z. (4)

UCLES

6 (i) Expand $(1-2x)^{\frac{1}{2}}$ in ascending powers of x up to and including the term in x^4 and state the range for which this expansion is valid. (6)

(ii) Use this expansion to deduce the square root of 0.8 correct to *four* decimal places. (4)

NICCEA

7 Given that $|x| < \frac{1}{4}$, write down the expansion of $(1-4x)^{-\frac{1}{2}}$ in ascending powers

of x up to and including the term in x^3. (4)

Hence obtain the coefficient of x^3 in the expansion of $\dfrac{(1-3x)}{\sqrt{(1-4x)}}$. (2)

AEB

8 The nth terms of two sequences are defined as follows:

(a) $t_n = 1-\dfrac{1}{n}$ (1)

(b) $u_n = 1-\dfrac{1}{u_{n-1}}$, where $u_1 = 2$. (3)

Decide in each case whether the sequence is convergent, divergent or oscillating or periodic, giving reasons for your answers.

Oxford

**REVISION
SUMMARY**

To find the **distance between two points**, use Pythagoras' Theorem where

$$d = \sqrt{(x_2 - x_1)^2 + (y_2 - y_1)^2}\,.$$

The **gradient** at a point on a line or a curve is the **rate of change of y in relation to x** at that point.

The **gradient of a line** is constant and it is often denoted by m where $m = \dfrac{y_2 - y_1}{x_2 - x_1}$.

The **mid-point** of the line joining (x_1, y_1) and (x_2, y_2) is $(\tfrac{1}{2}(x_1 + x_2), \tfrac{1}{2}(y_1 + y_2))$.

The most general format of the **equation of a line** is $ax + by + c = 0$. However the form that is often more useful is $y = mx + c$ where m is the gradient and c is the value of y when $x = 0$, i.e. the y-intercept. In this format it is easy to sketch lines provided that you remember the direction of the slope.

To find the equation of a line given the gradient m and a point (x_1, y_1) on the line,

- use $y = mx + c$, substituting the values for x_1 and y_1 to obtain c, or
- use $y - y_1 = m(x - x_1)$.

To find the equation of a line given two points (x_1, y_1) and (x_2, y_2) on the line, work out the gradient and then proceed as above.

Given two lines with gradients m_1 and m_2

- the lines are **parallel** if $m_1 = m_2$ (i.e. the gradients are the same),
- the lines are **perpendicular** if $m_1 \times m_2 = -1$ (i.e. the product of the gradients is -1).

Make sure that you are familiar with the general shape of the following **common curves**:

See also:
Functions,
Unit 4.

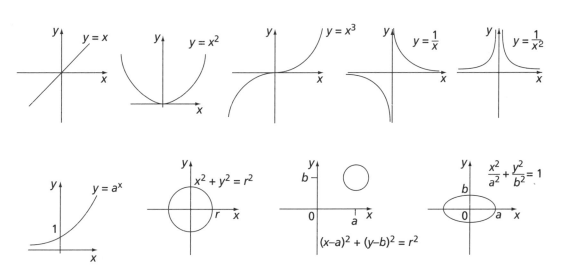

**If you need to
revise this
subject more
thoroughly,
see the relevant
topics in the
Letts A level
Mathematics
Study Guide.**

Parametric equations can be used to draw graphs. The x- and y-coordinates are each written in terms of a third variable, called a **parameter.**

For example, to draw $x = 4 + t$, $y = t^2$, find the coordinates by substituting various values of t. This process can be performed on a graphics calculator.

3 *Coordinate geometry*

1 The points *P, Q* and *R* have coordinates (2, 4), (7, –2) and (6, 2) respectively.

Find the equation of the straight line *l* which is perpendicular to the line *PQ* and which passes through the mid-point of *PR*. (5)

AEB

2 Find the equation of the straight line that passes through the points (3, –1) and (–2, 2), giving your answer in the form $ax + by + c = 0$. (3)

Hence find the coordinates of the points of intersection of the line and the *x*-axis. (2)

UCLES

3 The point *A* has coordinates (2, –5). The straight line $3x + 4y – 36 = 0$ cuts the *x*-axis at *B* and the *y*-axis at *C*. Find

(a) the equation of the line through *A* which is perpendicular to the line *BC*; (2)

(b) the perpendicular distance from *A* to the line *BC*; (3)

(c) the area of triangle *ABC*. (2)

Oxford

4 The line *l* has equation $2x – y – 1 = 0$. The line *m* passes through the point *A*(0, 4) and is perpendicular to the line *l*.

(a) Find an equation of *m* and show that the lines *l* and *m* intersect at the point *P*(2, 3).

The line *n* passes through the point *B*(3, 0) and is parallel to *m*.

(b) Find an equation of *n* and hence find the coordinates of the point *Q* where the lines *l* and *n* intersect.

(c) Prove that $AP = BQ = PQ$. (12)

ULEAC

5 *A, B* and *C* are the points (0, 2) (5, 7) and (12, 0) respectively.

(i) Find the lengths *AB, BC* and *CA* of the sides of the triangle *ABC*, and show that $AB^2 + BC^2 = CA^2$. (4)

Deduce the size of angle *ABC*. (1)

(ii) Find the gradients of the lines *AB* and *BC* and show how these can be used to confirm your answer in part (i) for the size of angle *ABC*. (3)

(iii) *M* is the mid-point of line *CA*. Show that $MA = MB$. (3)

(iv) Hence write down, but do not simplify, the equation of the circle through *A, B* and *C*. (3)

Oxford & Cambridge (MEI, specimen)

6 The curve *C* has parametric equations $x = at, y = \dfrac{a}{t}, t \in \mathbb{R}, t \neq 0$,

where *t* is a parameter and *a* is a positive constant.

(a) Sketch *C*.

(b) Find $\dfrac{dy}{dx}$ in terms of *t*.

The point *P* on *C* has parameter $t = 2$.

(c) Show that an equation of the normal to *C* at *P* is $2y = 8x – 15a$.

The normal meets *C* again at the point *Q*.

(d) Find the value of *t* at *Q*. (15)

ULEAC

A **function** may be thought of as a rule which assigns a *unique* value to each element of a given set.

$$x \longrightarrow \{ \text{function} \} \longrightarrow y$$

The assigned value, in this case *y*, is referred to as the **image** of *x* under the function and may be denoted by $f(x)$, read as '*f* of *x*'. The set of values on which the function acts is called the **domain** and the corresponding set of image values is called the **range**. If for each element *y* in the range, there is a *unique* value of *x* such that $f(x)=y$ then *f* is a **one-one** function. If for any element *y* in the range, there is more than one value of *x* satisfying $f(x)=y$ then *f* is **many-one**.

The function which adds 7 to the square of every real number, for example, might be written as

$$f(x) = x^2 + 7, \ x \in \mathbb{R} \quad \text{or alternatively as} \quad f : x \to x^2 + 7, \ x \in \mathbb{R} \qquad (f \text{ is many-one}).$$

The range of *f*, in this case, is given by $\{x : x \geq 7\}$. (Note: This is the same as $\{y : y \geq 7\}$).

In some cases, particular values must be omitted from the domain for the function to be valid.

For example, $f(x) = \dfrac{x+2}{x-3}$, $x \in \mathbb{R}$ $x \neq 3$ (division by zero is undefined).

If *f* and *g* are two functions then the **composite** function *fg* (sometimes written as $f \circ g$) is found by applying *g* first, followed by *f*, i.e. $fg(x) = f(g(x))$. It follows that the range of *g* becomes the domain of *f*.

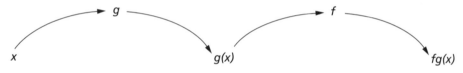

The *order* in which the functions are applied is very important. For example, if $f(x) = x^2$ and $g(x) = x - 1$, then $fg(x) = f(x-1) = (x-1)^2$ whereas $gf(x) = g(x^2) = x^2 - 1$.

If *g* is a function such that $gf(x) = x$ for all values *x* in the domain of *f*, then the effect of *g* is to *undo* what *f* has done and *g* is described as the **inverse** of *f*, denoted by f^{-1}. Only one-one functions have an inverse function, otherwise when the process is reversed the image may not be *unique*. One way to avoid this problem is to restrict the domain of the original function. For example, if $f(x) = \sin x$, where *x* can take all real values, then *f* does not have an inverse function. However, if the domain of *f* is restricted to $\{x : -90° \leq x \leq 90°\}$ then the inverse function exists.

The notation $\sin^{-1}(x)$ or arc sin *x* is used to denote this important function.

If you need to revise this subject more thoroughly, see the relevant topics in the *Letts* A level Mathematics Study Guide.

The graphs of some functions can be obtained by **transforming** the graph of a given function.

Function	Transformation of the graph of $y = f(x)$
$y = f^{-1}(x)$	Reflection in the line $y = x$.
$y = f(x) + a$	Translation described by $\begin{pmatrix} 0 \\ a \end{pmatrix}$
$y = af(x)$	One way stretch with scale factor *a* from $y = 0$.
$y = f(x-a)$	Translation described by $\begin{pmatrix} a \\ 0 \end{pmatrix}$
$y = f(ax)$	One-way stretch with scale factor $\frac{1}{a}$ from $x = 0$.

See also: Coordinate geometry, Unit 3.

4 Functions

1 The function f is defined by

$$f : x \mapsto \frac{1}{2-x} + 3, \qquad x \in \mathbb{R}, \ x \neq 2.$$

(a) Calculate $f(5)$ and $ff(5)$.

State the value of $k \ (k \neq 2)$ for which $ff(k)$ is not defined. (2)

(b) The inverse of f is f^{-1}. Find an expression for $f^{-1}(x)$ and state the domain of f^{-1}. (3)

AEB

2 $f(x) = 2x - 1$, $g(x) = 3 - 2x$ and $h(x) = \frac{1}{4}(5 - x)$.

(a) Find a formula for $k(x)$ where $k(x) = f(g(x))$. (2)

(b) Find a formula for $h(k(x))$. (2)

(c) What is the connection between the functions h and k? (1)

SEB

3 The functions f and g are defined by

$$f : x \mapsto 3x - 1, \quad x \in \mathbb{R}$$
$$g : x \mapsto x^2 + 1, \quad x \in \mathbb{R}$$

(a) Find the range of g.

(b) Determine the values of x for which $gf(x) = fg(x)$.

(c) Determine the values of x for which $|f(x)| = 8$.

The function $h : x \mapsto x^2 + 3x, \ x \in \mathbb{R}, \ x \geq q$, is one-one.

(d) Find the least value of q and sketch the graph of this function. (13)

ULEAC

4 The function f has domain the set of all non-zero real numbers, and is given by $f(x) = \frac{1}{x}$ for all x in this set. On a single diagram, sketch each of the following graphs, and indicate the geometrical relationships between them.

(i) $y = f(x)$,

(ii) $y = f(x+1)$,

(iii) $y = f(x+1) + 2$. (5)

Deduce, explaining your reasoning, the coordinates of the point about which the graph of $y = \dfrac{2x+3}{x+1}$ is symmetrical. (3)

UCLES

5 The function f with domain $\{x : x \geq 0\}$ is defined by $f(x) = 4x^2 - 1$.

(a) State the range of f and sketch the graph of f. (1)

(b) Explain why the inverse function f^{-1} exists and find $f^{-1}(x)$. (3)

(c) Given that g has domain $\{x : x \geq 0\}$ and is defined by $g(x) = \sqrt{(x+6)}$, solve the inequality

$$fg(x) \geq f(x).$$ (3)

Oxford

Angles are measured in **degrees** or **radians**. One complete revolution of 360° is equivalent to 2π radians and one radian is about 57°. Make sure that your calculator is set to the required mode, remembering in particular that for all calculus operations, you must be in radians. You should learn the most common conversions:

$\frac{\pi}{6}=30°$, $\frac{\pi}{4}=45°$, $\frac{\pi}{3}=60°$, $\frac{\pi}{2}=90°$, $\frac{2\pi}{3}=120°$, $\frac{3\pi}{4}=135°$, $\frac{5\pi}{6}=150°$, $\pi=180°$, $\frac{3\pi}{2}=270°$

When θ is measured in radians,
  arc length, $s=r\theta$

  area of sector $=\frac{1}{2}r^2\theta$.

You should be able to solve trigonometry problems in 2 or 3 dimensions, recalling the sin, cos and tan ratios in right-angled triangles and the following rules that apply in *any* triangle.

Sine Rule: $\dfrac{a}{\sin A}=\dfrac{b}{\sin B}=\dfrac{c}{\sin C}$

Cosine Rule: $a^2=b^2+c^2-2bc\cos A$

Area of $\triangle ABC=\frac{1}{2}ab\sin C$

Be aware of the **ambiguous case of the sine rule**. This arises because there are two solutions between 0° and 180° of the equation $\sin\theta=c$, for example $\sin\theta=0.5\Rightarrow\theta=30°$ or 150°. Sometimes one of the solutions can be eliminated (since the smallest angle is opposite the smallest side and the largest opposite the largest side), but there are occasions when both answers are possible.

It is useful to remember the following **special trigonometric ratios**, especially when you are required to give *exact* answers. You should also know the corresponding results for radians.

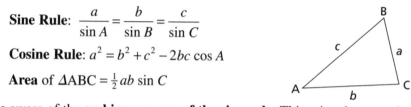

$\cos 60°=\sin 30°=\dfrac{1}{2}$

$\cos 30°=\sin 60°=\dfrac{\sqrt{3}}{2}$

$\tan 30°=\dfrac{1}{\sqrt{3}}$, $\tan 60°=\sqrt{3}$

$\sin 45°=\dfrac{1}{\sqrt{2}}$

$\cos 45°=\dfrac{1}{\sqrt{2}}$

$\tan 45°=1$

The definitions of the trigonometric ratios can be extended to include **any angle**, positive or negative, and you must make sure that you are familiar with the **graphs** of **sin θ, cos θ** and **tan θ**. They are all **periodic**, with $\sin\theta$ and $\cos\theta$ having period 360° (2π) and $\tan\theta$ having period 180° (π); $\sin\theta$ and $\cos\theta$ take values between −1 and 1; the range of $\tan\theta$ is unlimited.

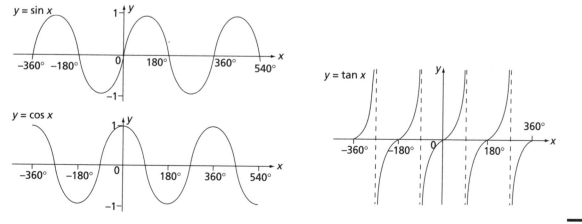

Remember that

$$\tan \theta = \frac{\sin \theta}{\cos \theta}, \qquad \operatorname{cosec} \theta = \frac{1}{\sin \theta}, \qquad \sec \theta = \frac{1}{\cos \theta}, \qquad \cot \theta = \frac{\cos \theta}{\sin \theta}.$$

We can see from the **periodic** and **symmetric properties** of the trigonometric functions that there are many solutions to equations such as $\sin \theta = 0.5$. Your calculator will give just one, in this case $\sin^{-1} 0.5 = 30°$ ($\pi/6$) known as the **principal value** of θ. For the sin and tan functions, the principal value lies between $-90°$ and $90°$ ($-\pi/2$ and $\pi/2$) and for the cos function, between $0°$ and $180°$ (0 and π). You will need to be proficient at working out other angles, usually in a specified range, once the principal value is known. All possible solutions can be summarised in the form of a **general solution**.

● To solve $\cos 2\theta = 0.5$ in the range $0 < \theta < 2\pi$, first find values of **2θ** in the range $0 < 2\theta < 4\pi$. These are $2\theta = \pi/3, 2\pi - \pi/3, 2\pi + \pi/3, 4\pi - \pi/3$, i.e $2\theta = \pi/3, 5\pi/3, 7\pi/3, 11\pi/3 \Rightarrow \theta = \pi/6, 5\pi/6, 7\pi/6, 11\pi/6$.

Learn the basic **Pythagorean identity** $\sin^2 \theta + \cos^2 \theta \equiv 1$, from which the following can be deduced: $\tan^2 \theta + 1 \equiv \sec^2 \theta$ and $1 + \cot^2 \theta \equiv \operatorname{cosec}^2 \theta$. These are often needed when proving identities or solving equations.

● To solve $6\cos^2 \theta + \sin \theta = 5$, substitute $1 - \sin^2 \theta$ for $\cos^2 \theta$ and solve the quadratic equation in $\sin \theta$ so formed.

In the **addition formula** a common mistake is to write $\sin (A + B)$ as $\sin A + \sin B$, so you must be very clear that $\sin (A + B) \equiv \sin A \cos B + \cos A \sin B$. Make sure that you are familiar with the expansions of $\sin (A \pm B)$, $\cos (A \pm B)$ and $\tan (A \pm B)$. In particular, note that $\tan (A - B)$ is useful when finding the angle between two lines.

An important application of the addition formulae is that it allows functions of the type $a \cos \theta \pm b \sin \theta$ to be written in the form $R \cos (\theta \mp \alpha)$ and $a \sin \theta \pm b \cos \theta$ to be written in the form $R \sin (\theta \pm \alpha)$, with $R = \sqrt{a^2 + b^2}$ and $\tan \alpha = b/a$. Note that the maximum and minimum values are given by $\pm R$.

● The maximum value of $f(\theta) = 3 \cos \theta + 4 \sin \theta$ is $\sqrt{3^2 + 4^2} = 5$ and the minimum value is -5.

**If you need to
revise this
subject more
thoroughly,
see the relevant
topics in the**
Letts **A level
Mathematics
Study Guide.**

● To sketch the curve, write it as $f(\theta) = 5 \cos (\theta - \alpha)$ where $\tan \alpha = \frac{4}{3} \Rightarrow \alpha = 53.1°$. Note that the curve is a transformation of $y = \cos \theta$, applying a one-way stretch with scale factor 5 from $y = 0$ (changing the amplitude to 5) and translating it by $53.1°$ to the right.

● To solve $3 \cos \theta + 4 \sin \theta = 2$, write it as $5 \cos (\theta - \alpha) = 2 \Rightarrow \cos (\theta - \alpha) = 0.4$, with $\alpha = 53.1°$. The advantage of this format is that θ appears only once in the equation.

Substituting $B = A$ in the addition formulae gives the **double angle** rules which are very important. Remember that $\sin 2A \equiv 2 \sin A \cos A$, $\tan 2A \equiv \dfrac{2 \tan A}{1 - \tan^2 A}$ and that $\cos 2A$ can be written in three different formats, where $\cos 2A \equiv 2\cos^2 A - 1 \equiv 1 - 2\sin^2 A \equiv \cos^2 A - \sin^2 A$. It is very useful to remember that $\cos^2 A \equiv \frac{1}{2} (1 + \cos 2A)$ and $\sin^2 A \equiv \frac{1}{2} (1 - \sin 2A)$.

● To solve $\sin 2x = \sqrt{3} \sin x$, write it as $2\sin x \cos x - \sqrt{3} \sin x = 0 \Rightarrow \sin x(2\cos x - \sqrt{3}) = 0$, so either $\sin x = 0$ or $\cos x = \sqrt{3}/2$. Do not be tempted to 'cancel' the factor of $\sin x$ which would result in the loss of the solutions relating to $\sin x = 0$.

For **small angles** θ, measured in radians, $\sin \theta \approx \theta$, $\tan \theta \approx \theta$, and $\cos \theta \approx 1 - \frac{1}{2}\theta^2$.

1  The diagram shows a sector of a circle, with centre O and radius r. The length of the arc is equal to half the perimeter of the sector. Find the area of the sector in terms of r.

(3)

UCLES

2 The diagram shows a triangle ABC in which $AB = 5$ cm, $AC = BC = 3$ cm. The circle, centre A, radius 3 cm, cuts AB at X; the circle, centre B, radius 3 cm, cuts AB at Y.

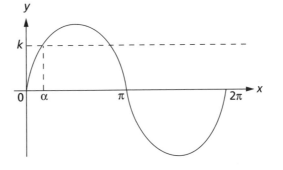

(a) Determine the size of the angle CAB, giving your answer in radians to four decimal places. (1)

(b) The region R, shaded in the diagram, is bounded by the arcs CX, CY and the straight line XY.

Calculate

(i) the length of the perimeter of R; (1)

(ii) the area of the sector ACX; (1)

(iii)the area of the region R. (2)

Oxford

3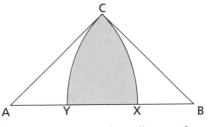

The diagram shows part of the graph of $y = \sin x$, where x is measured in radians, and values α on the x-axis and k on the y-axis such that $\sin \alpha = k$. Write down, in terms of α,

(i) a value of x between $\frac{1}{2}\pi$ and π such that $\sin x = k$, (1)

(ii) two values of x between 3π and 4π such than $\sin x = -k$. (2)

UCLES

4 (a) Find the values of $\cos x$ for which
$$6 \sin^2 x = 5 + \cos x.$$

(b) Find all the values of x in the interval $180° < x < 540°$ for which
$$6 \sin^2 x = 5 + \cos x.$$
(8)

ULEAC

5 Trigonometry

5 In triangle ABC, $AB = 8$ cm, $BC = 6$ cm and angle $B = 30°$.

(a) Find the length of AC. (3)

(b) Find the size of angle A. (3)

Oxford

6 The depth of water at the entrance to a harbour is y metres at time t hours after low tide. The value of y is given by

$$y = 10 - 3 \cos kt,$$

where k is a positive constant.

(a) Write down, or obtain, the depth of water in the harbour

    (i) at low tide; (1)

    (ii) at high tide. (1)

(b) Show by means of a sketch graph how y varies with t between two successive low tides. (2)

(c) Given that the time interval between a low tide and the next high tide is 6.20 hours, calculate, correct to two decimal places, the value of k. (2)

NEAB

7 Prove that $\sin 3\theta = 3 \sin \theta - 4 \sin^3 \theta$. (3)

Hence find all values of θ, for $0° \leq \theta \leq 360°$, which satisfy the equation $\sin 3\theta = 2 \sin \theta$. (5)

UCLES

8 (a) For the function $f(x) = \cos x° - \tan x°$, write down the values of x in the interval from -180 to 360 for which the function is undefined. (1)

(b) On the same axes, sketch the graphs of $y = \cos x°$ and $y = \tan x°$ for $-180 \leq x \leq 360$. (1)

(c) Use your sketch to determine the number of roots of the equations $f(x) = 0$ in the interval $-180 \leq x \leq 360$. (1)

(d) Find pairs of successive integers between which the roots lie. (2)

Oxford

9 Solve the equation $4 \tan^2 x + 12 \sec x + 1 = 0$, giving all solutions in degrees, to the nearest degree, in the interval $-180° < x < 180°$. (6)

AEB

10 The diagram shows the graph of the function $y = a + b\sin cx$ for $0 \leq x \leq \pi$.

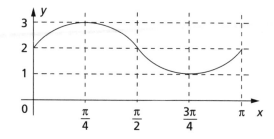

(a) Write down the values of a, b and c. (3)

(b) Find algebraically the values of x for which $y = 2·5$. (3)

SEB

The **gradient of the curve** $y = f(x)$ at the point $P(x, y)$ is given by the gradient of the **tangent** at P.

To find this *geometrically*, consider the gradients of a sequence of chords PQ such that as $Q \to P$, the chord $PQ \to$ the tangent at P.

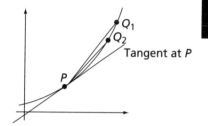

The gradient can be found *algebraically* from the

gradient function of $y = f(x)$, written $\dfrac{dy}{dx}$, $\dfrac{d}{dx}(f(x))$ or $f'(x)$

To obtain the gradient function from **first principles** use:

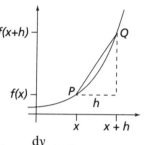

$$f'(x) = \lim_{h \to 0} \frac{f(x+h) - f(x)}{h}$$

The process of finding the gradient function is called **differentiation**, where $\dfrac{dy}{dx}$ is the **derivative**, or **differential**, of y with respect to x. It gives the **rate of change** of y with respect to x. If $\dfrac{dy}{dx} > 0$, the function is increasing and if $\dfrac{dy}{dx} < 0$, the function is decreasing.

When differentiating from first principles, we use the limit definition. More often, though, we quote and use standard results and it is likely that in an examination you will be given many of these in a reference booklet. Make sure that you are familiar with them, using the booklet only for a final check on accuracy, not for inspiration!

You will know that if $y = kx^n$, then $\dfrac{dy}{dx} = nkx^{n-1}$. This rule is easy to apply, but care must be taken with negative and fractional indices:

$$y = \frac{1}{\sqrt{x}} + \frac{4}{x^2} = x^{-1/2} + 4x^{-2}, \quad \frac{dy}{dx} = -\frac{1}{2}x^{-3/2} - 8x^{-3} = -\frac{1}{2x^{3/2}} - \frac{8}{x^3}$$

Some expressions can be simplified first before differentiating, such as

$$\frac{d}{dx}\left(\frac{(3x-1)(x+2)}{x}\right) = \frac{d}{dx}\left(\frac{3x^2 + 5x - 2}{x}\right) = \frac{d}{dx}\left(3x + 5 - 2x^{-1}\right) = 3 + 2x^{-2}$$

The **chain rule** is used to differentiate **composite** functions. This is sometimes known as differentiating **a function of a function**. If y is a function of u, and u is a function of x, then $\dfrac{dy}{dx} = \dfrac{dy}{du} \times \dfrac{du}{dx}$.

For example, to differentiate $y = (3x^2 + 5)^8$, let $y = u^8$ where $u = 3x^2 + 5$, then

$$\frac{dy}{du} = 8u^7 \text{ and } \frac{du}{dx} = 6x \text{ so } \frac{dy}{dx} = 8u^7 \times 6x = 48x(3x^2 + 5)^7$$

This technique is used when differentiating **implicitly**.

Note that $\dfrac{d}{dx}(y^2) = \dfrac{d}{dy}(y^2)\dfrac{dy}{dx} = 2y\dfrac{dy}{dx}$, so if $y^2 + 3x^2 - y = 4x$,

then $2y\dfrac{dy}{dx} + 6x - \dfrac{dy}{dx} = 4 \Rightarrow \dfrac{dy}{dx} = \dfrac{4 - 6x}{2y - 1}$

You should be familiar with these **standard results**:

$$\frac{d}{dx}(\sin ax) = a\cos ax \qquad \frac{d}{dx}\left(e^{ax}\right) = ae^{ax} \qquad \frac{d}{dx}(\ln(ax+b)) = \frac{a}{ax+b}$$

$$\frac{d}{dx}(\cos ax) = -a\sin ax \qquad \frac{d}{dx}\left(e^{f(x)}\right) = f'(x)e^{f(x)} \qquad \frac{d}{dx}(\ln f(x)) = \frac{f'(x)}{f(x)}$$

$$\frac{d}{dx}(\tan ax) = a\sec^2 ax$$

When expressions cannot be simplified, it may be necessary to use one of the following.

Product rule: **Quotient rule:**

$$\frac{d}{dx}(uv) = u\frac{dv}{dx} + v\frac{du}{dx} \qquad\qquad \frac{d}{dx}\left(\frac{u}{v}\right) = \frac{v\dfrac{du}{dx} - u\dfrac{dv}{dx}}{v^2}$$

If x and y are both functions of another variable, say t, then $\dfrac{dy}{dx} = \dfrac{dy}{dt} \times \dfrac{dt}{dx}$. This is called **parametric differentiation**.

Note that $\dfrac{dt}{dx} = \dfrac{1}{dx/dt}$. For example, if $x = t^3$, $y = t^2$, then

$$\frac{dx}{dt} = 3t^2 \Rightarrow \frac{dt}{dx} = \frac{1}{3t^2} \; ; \; \frac{dy}{dt} = 2t; \text{ therefore } \frac{dy}{dx} = 2t \times \frac{1}{3t^2} = \frac{2}{3t}.$$

To find the **equation of the tangent** at $P(x_1, y_1)$ on a curve, find the value of $\dfrac{dy}{dx}$ at P. Call this m, then use $y - y_1 = m(x - x_1)$. The gradient of the **normal** can be found from the gradient of the tangent, remembering that for two perpendicular lines, $m_1 \times m_2 = -1$.

A **stationary point** is one where $\dfrac{dy}{dx} = 0$. It could be one of the following:

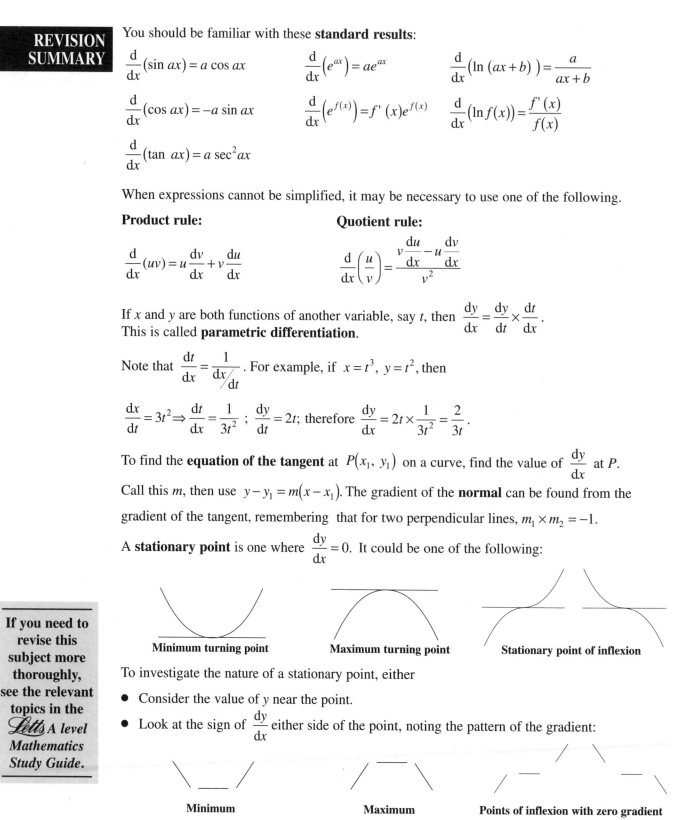

Minimum turning point   Maximum turning point   Stationary point of inflexion

To investigate the nature of a stationary point, either

If you need to revise this subject more thoroughly, see the relevant topics in the *Letts* A level Mathematics Study Guide.

- Consider the value of y near the point.
- Look at the sign of $\dfrac{dy}{dx}$ either side of the point, noting the pattern of the gradient:

Minimum   Maximum   Points of inflexion with zero gradient

- Consider the sign of the second differential $\dfrac{d}{dx}\left(\dfrac{dy}{dx}\right)$ written $\dfrac{d^2y}{dx^2}$.

If $\dfrac{dy}{dx} = 0$ and $\dfrac{d^2y}{dx^2} > 0$ there is a minimum turning point; if $\dfrac{dy}{dx} = 0$ and $\dfrac{d^2y}{dx^2} < 0$ there is a maximum turning point. Note that if the second differential is zero, then no conclusions can be drawn and one of the other methods must be used.

Letts

Q&A

1   Find the derivative, with respect to *x*, of

$$\frac{1}{x^3} + \cos\ 3x.$$

(4)

SEB

2   The radius *r* cm of a circular ink spot, *t* seconds after it first appears, is given by

$$r = \frac{1+4t}{2+t}.$$

Calculate

(a) the time taken for the radius to double its initial value;   (3)

(b) the rate of increase of the radius in cm s^{-1} when $t = 3$;   (5)

(c) the value to which *r* tends as *t* tends to infinity.   (2)

AEB

3   The curve with equation $ky = a^x$ passes through the points with coordinates (7,12) and (12, 7). Find, to 2 significant figures, the values of the constants *k* and *a*.

Using your values of *k* and *a*, find the value of $\dfrac{dy}{dx}$ at $x = 20,$ giving your answer to 1 decimal place.   (11)

ULEAC

4   Use differentiation to find the coordinates of the stationary points on the curve

$$y = x + \frac{4}{x},$$

and determine whether each stationary point is a maximum point or a minimum point.   (5)

Find the set of values of *x* for which *y* increases as *x* increases.   (3)

UCLES

5   Find the gradient of the tangent to the parabola $y = 4x - x^2$ at (0, 0).

Hence calculate the size of the angle between the line $y = x$ and this tangent.   (6)

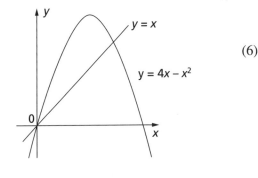

SEB

6   The parametric equations of a curve are given by

$$x = 160t - 6t^2,\ y = 80t - 8t^2.$$

Find the value of $\dfrac{dy}{dx}$ at each of the points on the curve where $y = 0.$ (5)

UCLES

23

QUESTIONS

7   In a medical treatment 500 milligrammes of a drug are administered to a patient. At time *t* hours after the drug is administered *X* milligrammes of the drug remain in the patient. The doctor has a mathematical model which states that

$$X = 500e^{-\frac{1}{5}t}.$$

(a) Find the value of *t*, correct to two decimal places, when *X* = 200. (3)

(b) (i) Express $\dfrac{\mathrm{d}X}{\mathrm{d}t}$ in terms of *t*. (2)

  (ii) Hence show that when *X* = 200 the rate of decrease of the amount of the drug remaining in the patient is 40 milligrammes per hour. (1)

NEAB

8   The graph of

$$y = x^3 + bx^2 + cx$$

is illustrated below.

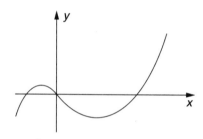

(a) Write down an expression for $\dfrac{\mathrm{d}y}{\mathrm{d}x}$. (2)

The curve has stationary points when *x* = −1 and *x* = 3.

(b) Show that *b* = −3 and calculate the value of *c*. (5)

(c) Hence find the local maximum and minimum values of *y*. (2)

(d) The graph is now translated by $\begin{pmatrix} 0 \\ d \end{pmatrix}$. Find the ranges of values for *d* such that the translated graph will have only one zero. (2)

NEAB (SMP 16–19)

9

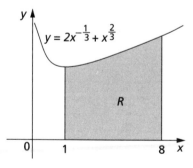

The figure shows a sketch of the curve with equation $y = 2x^{-\frac{1}{3}} + x^{\frac{2}{3}}$ for positive values of *x*.

(a) Find $\dfrac{\mathrm{d}y}{\mathrm{d}x}$ and hence show that at $x = 1$, $\dfrac{\mathrm{d}y}{\mathrm{d}x} = 0$.

The shaded region *R* is bounded by the curve, the *x*-axis and the lines *x* = 1 and *x* = 8.

(b) Determine the area of *R*. (12)

ULEAC

Integration 7

Integration can be thought of as the reverse process of differentiation. In order to integrate with confidence it is helpful to have a good knowledge of differentiation techniques, because when integrating a function, we often try to figure out what has been differentiated to give that function.

For **definite integration**, **limits** are given and a numerical value obtained. When there are no limits, the integration is known as **indefinite**, and it is necessary to include the **integration constant**, usually written c. To find the value of c, additional information is needed.

There are many **standard methods of integrating** and it is useful to adopt a systematic approach. Here are some techniques to consider:

For **powers of x**

when $n \neq -1$, $\int ax^n \, dx = \frac{a}{n+1} x^{n+1} + c$ and when $n = -1$, we have $\int \frac{a}{x} \, dx = a \ln|x| + c$.

Take care especially with negative and fractional indices when working out integrals.

The following integrals of **trigonometric functions** can be deduced from the differentials of $\sin ax$, $\cos ax$ and $\tan ax$, (see differentiation summary). Remember that angles are in radians.

$$\int \cos ax \, dx = \frac{1}{a} \sin ax + c \qquad \int \sin ax \, dx = -\frac{1}{a} \cos ax + c \qquad \int \sec^2 ax \, dx = \frac{1}{a} \tan ax + c$$

For example, $\int \left(3 \cos 2x - 4 \sec^2 3x\right) dx = \frac{3}{2} \sin + x - \frac{4}{3} \tan 3x = c$.

We can now integrate trigonometrical functions which can be written in one of these formats, such as

$$\int_0^{\pi/4} \tan^2 x \, dx = \int_0^{\pi/4} \left(\sec^2 x - 1\right) dx = \left[\tan x - x\right]_0^{\pi/4} = 1 - \frac{\pi}{4}.$$

To integrate **even powers of $\cos x$ and $\sin x$** use the double angle formula for cos, for example:

$$\int \sin^2 x \, dx = \int \frac{1}{2}(1 - \cos 2x) \, dx = \frac{1}{2}\left(x - \frac{1}{2} \sin 2x\right) + c$$

Consider whether a suitable **substitution** will help, for example to find

$$\int x(5x - 3)^8 \, dx, \text{ let } 5x - 3 = u \Rightarrow x = \frac{1}{5}(u + 3), \frac{dx}{du} = \frac{1}{5},$$

and the integral becomes $\int \frac{1}{5}(u + 3)u^8 \frac{1}{5} \, du = \frac{1}{25}\int \left(u^9 + 3u^8\right) du$.

For definite integrals, change the x limits to u limits when you make the substitution. If there are no limits, remember to change back to the original variable at the end.

Some integrals require a **trigonometric substitution**, for example, using $x = a \tan x$ we find that

$$\int \frac{1}{a^2 + x^2} \, dx = \frac{1}{a} \tan^{-1}\left(\frac{x}{a}\right) + c \text{ and using } x = a \sin x \text{ gives } \int \frac{1}{\sqrt{a^2 - x^2}} \, dx = \sin^{-1}\left(\frac{x}{a}\right) + c.$$

You can by-pass the substitution process if you **recognise an application of the chain rule**.

- $\dfrac{d}{dx}\left(x^4+2\right)^6 = 24x^3\left(x^4+2\right)^5 \Rightarrow \int x^3\left(x^4+2\right)^5 dx = \tfrac{1}{24}\left(x^4+2\right)^6+c$

- $\dfrac{d}{dx}\left(e^{-3x+2}\right) = -3e^{-3x+2} \Rightarrow \int e^{-3x+2}\, dx = -\tfrac{1}{3}e^{-3x+2}+c$

- $\dfrac{d}{dx}\left(\sin^4 x\right) = 4\sin^3 x\cos x \Rightarrow \int 5\sin^3 x\cos x\, dx = \tfrac{5}{4}\sin^4 x+c$

The last example illustrates the technique needed to integrate **odd powers of cos x or sin x**, for example to integrate $\cos^3 x$ write it as $\cos x\left(1-\sin^2 x\right) = \cos x - \cos x\sin^2 x$ and it can then be integrated directly by recognition.

So when **integrating by recognition**, make a guess at the integral, and then check it by differentiating, correcting the numerical factor if necessary. Look particularly for integrals of the type $\int f'(x)\left[f(x)\right]^n dx$.

For integrals in **quotient form**, look for a fraction in which the numerator is the differential of the denominator. This indicates that the integral is a **logarithmic function**, where $\int \dfrac{f'(x)}{f(x)}\, dx = \ln|f(x)|+c.$ It is often necessary to adjust the numerical factor, for example,

$$\int \frac{5x^2}{4x^3-1}\, dx = \tfrac{5}{12}\ln|4x^3-1|+c \qquad\qquad \int \tan x\, dx = \int \frac{\sin x}{\cos x}\, dx = -\ln|\cos x|+c = \ln|\sec x|+c$$

Check whether the denominator can be factorised and the function written in **partial fractions**.

$$\int \frac{x+13}{x^2+2x-15}\, dx = \int\left(\frac{2}{x-3}-\frac{1}{x+5}\right) dx = 2\ln|x-3|-\ln|x+5|+c.$$

When integrating a **product**, check whether it can be simplified, recognised or integrated by use of a substitution. If not, consider **integration by parts**, remembering that sometimes it is necessary to perform the procedure more than once. The formula is complicated to write out, but is straightforward to use in practice:

$$\int u\frac{dv}{dx}\, dx = uv - \int v\frac{du}{dx}\, dx.$$

To find $\int x\cos 2x\, dx$, let $u = x \Rightarrow \dfrac{du}{dx} = 1$ and let $\dfrac{dv}{dx} = \cos 2x \Rightarrow v = \tfrac{1}{2}\sin 2x$.

Therefore $\int x\cos 2x\, dx = x\tfrac{1}{2}\sin 2x - \int \tfrac{1}{2}\sin 2x(1)\, dx = \tfrac{1}{2}x\sin 2x + \tfrac{1}{4}\cos 2x + c.$

When integrating products which involve a power of x it is usual to take this as u, but note this particular case: write $\int x^n \ln x\, dx$ as $\int \ln x\left(x^n\right) dx$. A special application of this enables us to integrate $\ln x$, where

$$\int \ln x\, dx = \int \ln x\times 1\, dx = \ln x(x) - \int x\times\frac{1}{x}\, dx = x\ln x - x + c.$$

An important application of integration is in finding **areas and volumes**.

To find the **area** between the curve, the x-axis and the lines $x = a$, $x = b$:

$$\text{Area} = \int_a^b y\, dx$$

To find the **area** between the curve, the y-axis and the lines $y = c$, $y = d$:

$$\text{Area} = \int_c^d x\, dy$$

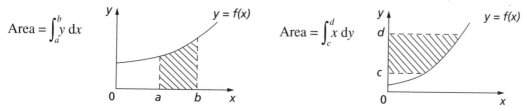

It is helpful to draw a sketch of the curve; this will enable you to spot any discontinuities. Remember also that some areas are negative and it may be necessary to integrate in sections.

The **area enclosed by two curves** is given by

$$\text{Area} = \int_{x_1}^{x_2} (y_1 - y_2)\, dx$$

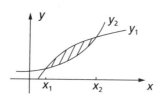

To find the **volume** obtained by rotating an area through one revolution:

About the x-axis: $\text{Volume} = \pi \int_a^b y^2\, dx$

About the y-axis: $\text{Volume} = \pi \int_c^d x^2\, dy$

You may be asked to form a **differential equation** from given information.

For example, if the rate of decay of m with t is proportional to m then $\dfrac{dm}{dt} \propto m$.

Since a rate of *decay* indicates a *negative* rate of change, we usually denote this by writing $\dfrac{dm}{dt} = -km$ with $k > 0$.

To find m, **separate the variables**, so

$$\int \frac{1}{m}\, dm = -\int k\, dt \implies \ln m = -kt + c.$$

Additional information enables the value of c, the integration constant, to be found. For example, if $m = m_o$ when $t = 0$, then $\ln m_o = c$ so

$$\ln m = -kt + \ln m_o \implies \ln\left(\frac{m}{m_o}\right) = -kt \implies m = m_o e^{-kt}.$$

Note that further information is required in order to find k, the proportionality constant.

1 (a) Differentiate $\left(1+x^3\right)^{\frac{1}{2}}$ with respect to x. (3)

(b) Use the result from (a), or an appropriate substitution, to find the value of

$$\int_0^2 \frac{x^2}{\sqrt{1+x^3}}\,\mathrm{d}x.$$

(4)
AEB

2 Curves C_1 and C_2 have equations $y=\dfrac{1}{x}$ and $y=kx^2$ respectively, where k is a constant.

The curves intersect at the point P, whose x-coordinate is $\frac{1}{2}$.

(a) Determine the value of k.

(b) Find the gradient of C_1 at P.

(c) Calculate the area of the finite region bounded by C_1, C_2, the x-axis and the line $x = 2$, giving your answer to 2 decimal places. (15)

ULEAC

3 (a) Sketch the area represented by the integral $\displaystyle\int_2^5 x^2\,\mathrm{d}x$ and calculate the value of the integral, showing your working. (2)

(b) Use your result from (a) to calculate the value of $\displaystyle\int_4^{25} \sqrt{y}\,\mathrm{d}y$. (2)

Note that you will obtain no marks unless you use your result from (a).

(c) Calculate the volume obtained by revolving the part of the graph of $y=\sqrt{x+1}$ between $x = 3$ and $x = 5$, through $360°$ around the x-axis, showing your working. (2)

Oxford (Nuffield)

4 (a) Differentiate $x \ln x$ with respect to x and hence show that the curve $y = x \ln x\ (x>0)$ has a minimum point for $x = \frac{1}{e}$. State the corresponding y coordinate. (5)

(b) Give a reason why the curve has no point of inflection. (1)

(c) Sketch the curve for $x \geq \frac{1}{e}$. (1))

(d) Find the point A where the curve meets the line $y = x$ and find the area enclosed by the line OA, the curve and the x-axis. (5)

NEAB

5 Find

(a) $\displaystyle\int x \cos x\,\mathrm{d}x,$

(b) $\displaystyle\int \cos^2 y\,\mathrm{d}y.$

Hence find the general solution of the differential equation

$$\frac{\mathrm{d}y}{\mathrm{d}x} = x \cos x \sec^2 2y, \quad 0 < y < \frac{\pi}{4}.$$

(12)
ULEAC

6 Radium is a radioactive substance. You can model its decay by the differential equation
$\frac{dR}{dt} = -kR$ where t is the time in years, R is the amount of radium in grams present at time t, and k is a positive constant.

Suppose that when $t = 0$, 10 g of radium is present.

(a) Solve the equation $\frac{dR}{dt} = -kR$ to find R in terms of t and k. (2)

(b) It is known that the amount of radium will have halved after about 1600 years. Use this information to show that $k = \frac{\ln 2}{1600}$. (2)

(c) According to this model, how many grams of radium will be left after 100 years? (2)

Oxford (Nuffield)

7 A circular patch of oil on the surface of water has radius r metres at time t minutes. When $t = 0$, $r = 1$ and when $t = 10$, $r = 2$. It is desired to predict the value T of t when $r = 4$.

(i) In a simple model the rate of increase of r is taken to be a constant. Find T for this model. (3)

(ii) In a more refined model, the rate of increase of r is taken to be proportional to $\frac{1}{r}$. Express this statement as a differential equation, and find the general solution. (4)

Find T for this second model. (4)

UCLES

8 Use the substitution $x = 2\cos\theta$ or otherwise, to evaluate
$$\int_1^{\sqrt{2}} \frac{1}{x^2\sqrt{4-x^2}}\,dx$$
giving your answer in surd form. (5)

Oxford

8 Numerical methods

Numerical methods provide us with an alternative approach to solving equations that may be difficult, or impossible, to solve algebraically. The use of graphics calculators, in particular, enables us to apply these methods with great efficiency.

Consider the equation $e^x - 4x - 3 = 0$. Sketching the graph of $y = e^x - 4x - 3$ provides important information about the nature of the solutions and their approximate values.

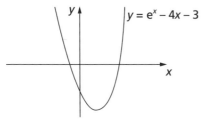

The **trace** function of a graphics calculator reveals the approximate solutions as –0.6 and 2.6.

Numerical methods may now be employed to refine these estimates to a high degree of accuracy.

Taking $f(x) = e^x - 4x - 3$, a **systematic search** is carried out by evaluating $f(x)$ at points near to a root, and looking for a change of sign. In this case, for example, $f(2.5) = -0.8... < 0$ and $f(2.6) = 0.06... > 0$, which tells us that the solution lies between 2.5 and 2.6. The table below shows how the process may be continued. The same approach is used to find the root near –0.6.

$x : f(x) < 0$	$x : f(x) > 0$	Solution in interval
2.5	2.6	(2.5, 2.6)
2.58		(2.58, 2.6)
2.59		(2.59, 2.6)
	2.595	(2.59, 2.595)
	2.594	(2.59, 2.594)
2.593		(2.593, 2.594)
	2.5935	(2.593, 2.5935)

A graphics calculator key sequence such as

$? \to X : e^x - 4X - 3$

allows $f(x)$ to be calculated, for different trial values, with the minimum of effort.

Since it has now been established that a root lies between 2.593 and 2.5935 it follows that its value may now be stated as 2.593 correct to 3 decimal places.

In essence, the **interval bisection** method is the same as the systematic search except that, at each stage, $f(x)$ is evaluated at the mid-point of the interval in which the root is known to lie.

The equation $e^x - 4x - 3 = 0$ can be rearranged as $x = \dfrac{e^x - 3}{4}$ to give the **iteration** formula $x_{n+1} = \dfrac{e^{x_n} - 3}{4}$.

If you need to revise this subject more thoroughly, see the relevant topics in the *Letts* A level Mathematics Study Guide.

The graph illustrates the effect of using this iteration formula from different starting points. The iteration always *fails* to converge to the upper root. However, it will converge to the lower root from any starting point less than the upper root.

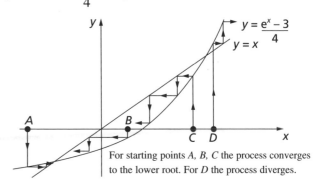

For starting points A, B, C the process converges to the lower root. For D the process diverges.

A different rearrangement of the original equation produces the iteration formula $x_{n+1} = \ln(4x_n + 3)$, which converges to the upper root from any starting point greater than the lower root. It follows that this re-arrangement fails to converge to the lower root.

Another fixed point iterative method, for the solution of $f(x) = 0$, is the **Newton-Raphson** method given by $x_{n+1} = x_n - \dfrac{f(x_n)}{f'(x_n)}$. For the given equation, this gives $x_{n+1} = x_n - \dfrac{e^{x_n} - 4x_n - 3}{e^{x_n} - 4}$ which can converge rapidly to *both* roots depending on the starting value.

1 The sequence given by the iteration formula

$$x_{n+1} = 2(1 + e^{-x_n}),$$

with $x_1 = 0$, converges to α. Find α correct to 3 decimal places, and state an equation of which α is a root. (4)

UCLES

2 (a) Show that the equation $x^3 + 3x^2 - 7 = 0$ may be rearranged into the form $x = \sqrt{\dfrac{a}{x+b}}$, and state the values of a and b.

(b) Hence, using the iteration formula

$$x_{n+1} = \sqrt{\dfrac{a}{x_n + b}}$$

with $x_0 = 2$ together with your values of a and b, find the approximate solution x_4 of the equation, giving your answer to an appropriate degree of accuracy. Show your intermediate answers and explain why the degree of accuracy you have chosen for x_4 is appropriate. (7)

ULEAC

3 Show that the equation $x^3 - x^2 - 2 = 0$ has a root α which lies between 1 and 2. (2)

(a) Using 1.5 as a first approximation for α, use the Newton-Raphson method once to obtain a second approximation for α, giving your answer to 3 decimal places. (4)

(b) Show that the equation $x^3 - x^2 - 2 = 0$ can be arranged in the form $x = \sqrt[3]{(f(x))}$ where $f(x)$ is a quadratic function.

Use an iteration of the form $x_{n+1} = g(x_n)$ based on this rearrangement and with $x_1 = 1.5$ to find x_2 and x_3, giving your answers to 3 decimal places. (5)

AEB

4 The function f is defined by

$$f(x) = e^x - 5x, \quad x \in \mathbb{R}$$

(a) Determine $f'(x)$.

(b) Find the value of x for which $f'(x) = 0$ giving your answer to 2 decimal places.

(c) Show, by calculation, that there is a root α of the equation $f(x) = 0$ such that $0.2 < \alpha < 0.3$.

(d) Determine the integer p such that the other root β of the equation $f(x) = 0$ lies in the interval

$$\frac{p}{10} < \beta < \frac{p+1}{10}.$$

(12)

ULEAC

5 On a single diagram, sketch the graphs of $y = \ln(10x)$ and $y = \dfrac{6}{x}$, and explain how you can

deduce that the equation $\ln(10x) = \dfrac{6}{x}$ has exactly one real root. (3)

Given that the root is close to 2, use the iteration

$$x_{n+1} = \frac{6}{\ln(10x_n)}$$

to evaluate the root correct to three decimal places. (2)

The same equation may be written in the form $x\ln(10x) - 6 = 0$. Taking $f(x)$ to be $x\ln(10x) - 6$, find $f'(x)$, and show that the Newton-Raphson iteration for the root of $f(x) = 0$ may be simplified to the form

$$x_{n+1} = \frac{x_n + 6}{1 + \ln(10x_n)}.$$

(5)

UCLES

6 The chord AB of a circle subtends an angle θ radians at the centre O of the circle of radius r, as shown in the diagram.

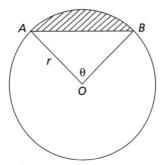

(a) Find an expression for the shaded area, in terms of r and θ. (3)

(b) Given that this shaded area is $\frac{1}{6}$ of the area of the circle, show that θ is given by

$$\sin\theta = \theta - \frac{\pi}{3}.$$

(1)

(c) By sketching the graphs of $y = \sin\theta$ and $y = \theta - \dfrac{\pi}{3}$ on the same diagram, verify that

$\theta = 2$ is an approximate solution of the equation $\sin\theta = \theta - \dfrac{\pi}{3}$. (2)

(d) Find a better approximation for θ using one application of Newton's rule. (3)

NEAB

When asked to suggest suitable ways of **collecting data**, bear in mind the purpose for which they are to be used and how best to **organise** them so that they will be easy to assimilate and interpret. You should be familiar with **methods of presentation** such as pie charts, histograms, frequency graphs, cumulative frequency graphs and scatter diagrams. You may be asked to discuss advantages and/or disadvantages of particular representations.

In a **histogram**, the area of a bar represents the frequency. Problems arise when intervals are of unequal widths, in which case it is easiest to use the **frequency density** of an interval as the height of the bar, where frequency density $= \dfrac{\text{frequency}}{\text{interval width}}$. A **frequency graph** is formed by plotting the frequency density against the **mid-points** of the intervals and joining the points with straight lines (frequency polygon) or a smooth curve.

Cumulative frequency represents a running total, and when drawing **cumulative frequency graphs**, remember to plot the cumulative frequency against the **upper class boundary** of an interval.

The measures of **central tendency** (sometimes called measures of location) are the three **averages**: mode, mean and median. For raw data, the **mode** is the value that occurs most often. When data have been grouped, the **modal class** is the one with the greatest frequency density. The **mean** is the arithmetic average and it takes account of every reading. Often you will use the statistical functions on your calculator, but you should be familiar with the formulae for the mean, remembering that for grouped data, the mid-point of the interval is used for x.
For raw data, $\bar{x} = \dfrac{\Sigma x}{n}$ and for data in a frequency distribution, $\bar{x} = \dfrac{\Sigma fx}{\Sigma f}$.

The **median** is the middle value when the data are arranged in order; for grouped data, find the $\frac{1}{2}n$th value either from a cumulative frequency curve or by **linear interpolation**. The median is a useful average, especially when the distribution contains extreme values.

The **standard deviation** gives a useful **measure of dispersion**, or **spread**, because it takes account of every reading. Note that in most distributions the bulk of the readings lie within two standard deviations of the mean. You will probably use your calculator to find it, but make sure that you can use the formula if necessary, where $s = \sqrt{\dfrac{\Sigma f(x - \bar{x})^2}{\Sigma f}}$ or $s = \sqrt{\dfrac{\Sigma fx^2}{\Sigma f} - \bar{x}^2}$.

Other measures of spread include the complete **range** (the difference between the highest and lowest readings) and the range of the middle half of the readings, known as the **interquartile range**. This is given by the difference between the upper quartile and the lower quartile, where the upper quartile is the $\frac{3}{4}n$th value and the lower quartile is the $\frac{1}{4}n$th value. A concise way of representing both of these ranges is on a **box and whisker diagram**, from which it is easy to assess the **skewness** of the distribution.

**If you need to
revise this
subject more
thoroughly,
see the relevant
topics in the
Letts A level
Mathematics
Study Guide.**

Probability: Two events, A and B, are **mutually exclusive** if the occurrence of one of them excludes the occurrence of the other, in which case $P(A \text{ or } B) = P(A) + P(B)$.

Two events, A and B, are **independent** if the outcome of one event does not affect the outcome of the other, in which case $P(A \text{ and } B) = P(A) \times P(B)$.

These two formulae are special cases of the following rules for **any two events**, A and B:
  (i) $P(A \text{ or } B) = P(A) + P(B) - P(A \text{ and } B)$, where $P(A \text{ or } B)$ means $P(A \text{ or } B \text{ or both})$;
  (ii) $P(A \text{ and } B) = P(A) \times P(B|A)$, where $P(B|A)$ is the **conditional probability** that B occurs, given that A occurs. Note that when the events are independent, $P(B|A) = P(B)$.

When finding probabilities, it is often helpful to use **possibility spaces** or **tree diagrams**.

1 An angler made a record of the weights (in lb) of the 200 fish he caught during one year. These are summarised in the table.

Weight of fish, lb (mid-class value)	0.5	1.25	1.75	2.25	2.75	3.5	4.5	5.5	7.0	10.5
Number of fish in the class	21	32	33	24	18	21	16	12	11	12
Class width	1	0.5	0.5	0.5	0.5	1	1	1	2	5

Using the information supplied in the table:

(a) calculate suitable frequency densities and, on graph paper, construct a histogram of the data; (4)

(b) calculate estimates of the mean and standard deviation of the weights of the fish. (3)

Oxford

2 The distribution of speeds of a sample of 250 vehicles on an inner city road is summarised in the following table. The road is subject to a 30 mph speed limit.

Speed, v, in mph	Number of vehicles
$10 < v \leq 20$	18
$20 < v \leq 25$	26
$25 < v \leq 30$	90
$30 < v \leq 35$	58
$35 < v \leq 40$	32
$40 < v \leq 50$	18
$50 < v \leq 70$	8

(a) (i) Represent these data by a cumulative frequency diagram (5)

    (ii) Estimate the median, lower quartile and upper quartile speeds of the vehicles. (3)

(b) (i) Use your estimates obtained in part (a) to draw an approximate box plot. (2)

    (ii) Comment briefly on the extent to which vehicles break the speed limit based on the evidence of this diagram. (2)

NEAB

3 The following information appears in the Annual Report 1992 of Eurotunnel PLC.

Size of shareholding	Number of shareholders
1 – 99	133 853
100 – 499	347 495
500 – 999	79 087
1 000 – 1 499	31 638
1 500 – 2 499	27 547
2 500 – 4 999	10 655
5 000 – 9 999	3 842
10 000 – 49 999	2 188
50 000 – 99 999	283
100 000 – 249 999	216
250 000 – 499 999	82
500 000 – 999 999	54
1 000 000 and over	50
Total	636 990

(i) Explain briefly what difficulties would arise in attempting to

    (a) represent the data by means of an accurate and easily comprehensible diagram,   (2)

    (b) find the mean size of shareholding per shareholder.   (1)

(ii) Estimate the median size of shareholding.   (3)

UCLES

4   Two events A and B are such that $p(A) = 0.4$, $p(B) = 0.7$, $p(A \text{ or } B) = 0.8$.

Calculate

(a) $p(A \text{ and } B)$;   (2)

(b) the conditional probability $p(A|B)$.   (2)

AEB

5   Doctors estimate that three people in every thousand of the population are infected by a particular virus. A test has been devised which is not perfect, but gives a positive result for 95% of those who have the virus. It also gives a positive result for 2% of those who do not have the virus.
Suppose that someone selected at random takes the test and that it gives a positive result. Calculate the probability that this person really has the virus.   (8)

Oxford (Nuffield)

6   A player has two dice, which are indistinguishable in appearance. One die is fair, so that the probability of getting a six on any throw is $\frac{1}{6}$, and one is biased in such a way that the

probability of getting a six on any throw is $\frac{1}{3}$.

(i) The player chooses one of the dice at random and throws it once.

    (a) Find the probability that a six is thrown.   (2)

    (b) Show that the conditional probability that the die is the biased one, given that a six is thrown, is $\frac{2}{3}$.   (2)

(ii) The player chooses one of the dice at random and throws it twice.

    (a) Show that the probability that two sixes are thrown is $\frac{5}{72}$.   (3)

    (b) Find the conditional probability that the die is the biased one, given that two sixes are thrown.   (2)

(iii)The player chooses one of the dice at random and throws it n times.
    Show that the conditional probability that the die is the biased one, given that

    n sixes are thrown, is $\dfrac{2^n}{2^n + 1}$.   (4)

UCLES

10 *Vectors*

Vector addition and subtraction

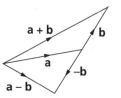

Multiplication by a scalar

In the usual notation, the letter O is used to denote the position of some fixed reference point called the **origin**. If P is some other point, then the vector from O to P (written as $\overrightarrow{OP}$ or **OP**) is referred to as the **position vector** of P.

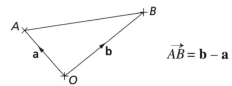

$$\overrightarrow{AB} = \mathbf{b} - \mathbf{a}$$

The position vector of a point dividing a line in a given ratio is given by

$$\overrightarrow{OP} = \frac{\alpha\mathbf{a} + \beta\mathbf{b}}{\alpha + \beta}$$

The **vector equation of a line** is in the form $\mathbf{r} = \mathbf{a} + t\mathbf{b}$ where $\mathbf{a}$ is the position vector of some fixed point on the line, $\mathbf{b}$ is a constant vector parallel to the line (often described as the directon vector) and t is a scalar parameter. As t changes, $\mathbf{r}$ represents the general position vector of a point on the line.

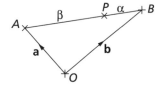

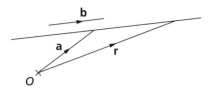

If the coordinates of P are (x, y, z) then the position vector $\overrightarrow{OP}$ may be written as $\begin{pmatrix} x \\ y \\ z \end{pmatrix}$ or as $x\mathbf{i} + y\mathbf{j} + z\mathbf{k}$.

If you need to revise this subject more thoroughly, see the relevant topics in the *Letts* **A level Mathematics Study Guide.**

The distance OP is given by the **magnitude** of $\overrightarrow{OP}$ which is $\left|\overrightarrow{OP}\right| = \sqrt{x^2 + y^2 + z^2}$.

The **scalar product a.b** is so-called because its value is a scalar. It is also known as the **dot product** because of the notation used.

(Note that $\mathbf{a} \times \mathbf{b}$ represents a *different* form of product in which the result is a vector).

$$\mathbf{a}.\mathbf{b} = |\mathbf{a}||\mathbf{b}|\cos\theta$$

A *very important* special case is that if $\mathbf{a}$ and $\mathbf{b}$ are perpendicular then $\mathbf{a}.\mathbf{b} = 0$. For example $\mathbf{i}.\mathbf{j} = \mathbf{i}.\mathbf{k} = \mathbf{j}.\mathbf{k} = 0$. Note also that $\mathbf{i}.\mathbf{i} = 1$, $\mathbf{j}.\mathbf{j} = 1$, $\mathbf{k}.\mathbf{k} = 1$. These results may be used to establish a very simple method of calculating $\mathbf{a}.\mathbf{b}$ when $\mathbf{a}$ and $\mathbf{b}$ are in component form.
If $\mathbf{a} = a_1\mathbf{i} + a_2\mathbf{j} + a_3\mathbf{k}$ and $\mathbf{b} = b_1\mathbf{i} + b_2\mathbf{j} + b_3\mathbf{k}$ then $\mathbf{a}.\mathbf{b} = a_1b_1 + a_2b_2 + a_3b_3$.

The angle between two vectors may then be given by $\theta = \cos^{-1}\left(\dfrac{\mathbf{a}.\mathbf{b}}{|\mathbf{a}||\mathbf{b}|}\right)$. To find the angle between two lines, find the angle between their direction vectors.

1 The vectors **a**, **b** and **c** are defined as follows:

$$\mathbf{a} = 2\mathbf{i} - \mathbf{k}, \quad \mathbf{b} = \mathbf{i} + 2\mathbf{j} + \mathbf{k}, \quad \mathbf{c} = -\mathbf{j} + \mathbf{k}.$$

(a) Evaluate **a.b** + **a.c**. (3)

(b) From your answer to part (a), make a deduction about the vector **b** + **c**. (2)

SEB

2

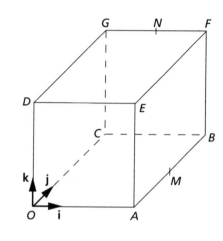

In the diagram *OABCDEFG* is a cube in which the length of each edge is 2 units. Unit vectors **i**, **j**, **k** are parallel to $\overrightarrow{OA}$, $\overrightarrow{OC}$, $\overrightarrow{OD}$ respectively. The mid-points of *AB* and *FG* are *M* and *N* respectively.

(i) Express each of the vectors $\overrightarrow{ON}$ and $\overrightarrow{MG}$ in terms of **i**, **j** and **k**. (3)

(ii) Show that the acute angle between the directions of $\overrightarrow{ON}$ and $\overrightarrow{MG}$ is 63.6°, correct to the nearest 0.1°. (5)

UCLES

3 Vectors **r** and **s** are given by

$$\mathbf{r} = \lambda\mathbf{i} + (2\lambda - 1)\mathbf{j} - \mathbf{k},$$

$$\mathbf{s} = (1 - \lambda)\mathbf{i} + 3\lambda\mathbf{j} + (4\lambda - 1)\mathbf{k},$$

where λ is a scalar.

(a) Find the values of λ for which **r** and **s** are perpendicular.

When $\lambda = 2$, **r** and **s** are the position vectors of the points *A* and *B* respectively, referred to an origin *O*.

(b) Find $\overrightarrow{AB}$.

(c) Use a scalar product to find the size of angle *BAO*, giving your answer to the nearest degree. (14)

ULEAC

4 The line l has vector equation $\mathbf{r} = 2\mathbf{i} + s(\mathbf{i} + 3\mathbf{j} + 4\mathbf{k})$.

    (a) (i) Show that the line l intersects the line with equation $\mathbf{r} = \mathbf{k} + t(\mathbf{i} + \mathbf{j} + \mathbf{k})$ and determine the position vector of the point of intersection. (5)

       (ii) Calculate the acute angle, to the nearest degree, between these two lines. (4)

    (b) Find the position vectors of the points on l which are exactly $5\sqrt{10}$ units from the origin. (5)

    (c) Determine the position vector of the point on l which is closest to the point with position vector $6\mathbf{i} - \mathbf{j} + 3\mathbf{k}$. (6)

Oxford

5 (a)

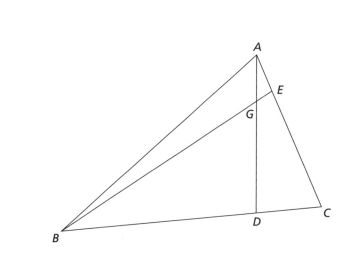

The points A, B, C, have position vectors $\mathbf{a}$, $\mathbf{b}$, $\mathbf{c}$ relative to an origin O. The points D, E lie on BC, CA respectively such that $BD : DC = 2 : 1$ and $CE : EA = 3 : 1$. The lines AD and BE meet at G.

    (i) Find the position vectors of D and E.

    (ii) Show that the position vector of G is

$$\frac{2}{3}\mathbf{a} + \frac{1}{9}\mathbf{b} + \frac{2}{9}\mathbf{c}.$$

    (iii) The line CG meets AB at F. Find the position vector of F in terms of $\mathbf{a}$ and $\mathbf{b}$. (10)

    (b) *OPQR* is a parallelogram and $\mathbf{OP} = \mathbf{p}$, $\mathbf{OR} = \mathbf{r}$.

       (i) Write down expressions in terms of $\mathbf{p}$ and $\mathbf{r}$ for $\mathbf{OQ}$ and $\mathbf{PR}$.

       (ii) Given that $OP = OR$, show by considering the scalar product of $\mathbf{OQ}$ and $\mathbf{PR}$ that OQ and PR are perpendicular. (5)

WJEC

The process of formulating a problem in mathematical terms, in preparation for further analysis, is referred to as **setting up a model**. In principle, the solution of the mathematical problem will then yield information about the original problem that would not have been evident without such an analysis.

A first step in setting up a model is to make a number of simplifying assumptions with a view to producing a workable representation of the problem. Typically, this may mean regarding any strings as light and inextensible, pulleys as light and frictionless, and projectiles as particles that move without air resistance.

The complete modelling process involves taking steps to ensure the validity of the results produced, and provides some measure of the reliability of any predictions based on the model.

Newton's Laws of Motion provide us with a clear set of rules that can be used to analyse the effect of **forces** within a system. This often enables us to make the all important link between the original problem and its mathematical representation through equations, as required in the modelling process.

The laws may be stated as:
❶ Every body continues in a state of rest or uniform motion in a straight line unless acted upon by an external force.
❷ The resultant force acting on a body is equal to its rate of change of momentum. ($\mathbf{F} = m\mathbf{a}$)
❸ Every force has an equal and opposite reaction.

The **moment** of a force is a measure of its turning effect, and is found by multiplying the magnitude of the force by its perpendicular distance from the axis of rotation. The combined turning effect of a number of **coplanar forces** about an axis is given by the algebraic sum of the clockwise and anti-clockwise moments i.e. one direction is taken as positive and the other as negative. A body is said to be in **equilibrium** if the resultant force acting on it is zero and the combined turning effect about any axis is also zero. Small bodies are often modelled by particles in which case the turning effect of the forces is neglected.

The diagram below shows how a force may be **resolved** into two perpendicular **components**.

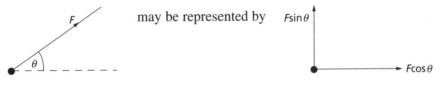

An important example concerns the resolution of the weight of an object, on an inclined plane, into components parallel to the plane and perpendicular to the plane. The diagrams below show the forces acting on an object of mass m in equilibrium on a plane inclined at angle θ to the horizontal. R is the **reaction force** exerted by the plane on the object and F is a force due to **friction**.

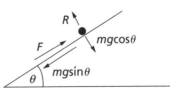

These components may be combined to give
Perpendicular to the plane $R - mg \cos \theta = 0$ (1)   (Newton's Third Law).
Parallel to the plane $mg \sin \theta - F = 0$ (2)   (Since the object is in equilibrium).
Note that in the case where $mg \sin \theta - F > 0$, the object will have an acceleration a down the plane given by $mg \sin \theta - F = ma$. In the case where the object is in **limiting equilibrium** i.e. it is on the point of slipping, the frictional force is given by $F = \mu R$ where μ is called the **coefficient of friction**. Substituting for R from (1) would give $F = \mu mg \cos \theta$.

A **vector** quantity has both magnitude and direction, whereas a **scalar** quantity has only magnitude. **Force** is therefore a vector quantity and the method of combining forces in their component form, described above, corresponds to vector addition.

Velocity is a vector quantity and **speed** is the corresponding scalar quantity. In the same way, **distance** is a scalar and **displacement** the corresponding vector.

> Vectors are dealt with in more detail in Unit 10.

The distinction between vectors and scalars is extremely important. For example, a consequence of Newton's First Law is that an object moving, even with *constant speed*, in a circular path must be subject to a force. This is consistent with Newton's Second Law since the changing direction of the motion corresponds to a *change in velocity* and subsequent non-zero **acceleration**.

The study of the motion of an object, or system of objects, without consideration of the physical laws that govern the motion is known as **kinematics**. In solving a kinematics problem, the assumption is made that an appropriate mathematical model has already been set up and that the focus of attention is on the details of the motion.

Displacement, velocity and acceleration are all related.	The formulae for *constant* acceleration provide important tools for the solution of some problems in kinematics.
$$\mathbf{s} = \int \mathbf{v}\, dt \qquad \mathbf{v} = \frac{d\mathbf{s}}{dt} = \int \mathbf{a}\, dt \qquad \mathbf{a} = \frac{d\mathbf{v}}{dt} = \frac{d^2\mathbf{s}}{dt^2}$$	
$\longrightarrow$ *differentiate* $\longrightarrow$	$\mathbf{v} = \mathbf{u} + \mathbf{a}t$
Displacement　　Velocity　　Acceleration	$\mathbf{s} = \mathbf{u}t + \frac{1}{2}\mathbf{a}t^2$
$\longleftarrow$ *integrate* $\longleftarrow$	$\mathbf{s} = \frac{1}{2}(\mathbf{u} + \mathbf{v})t$
Using the chain rule we also have $a = v\dfrac{dv}{ds}$.	$v^2 = u^2 + 2as$

The **S.I.** (Systeme International d'Unites) is a *coherent* system in which the unit of time is the second (s), the unit of distance is the metre (m), the unit of mass is the kilogram (kg), the unit of force is the Newton (N), the unit of energy is the Joule (J) and the unit of power is the watt (W). For example, the force required to make a mass of 1 kg accelerate at the rate of 1 ms^{-2} is 1 N.

In the usual model for **projectile problems,** at this level, air resistance is ignored and so the *only* force taken to be acting on an object in flight is its own **weight**. The model also assumes that the value of *g*, the acceleration due to gravity, is constant at 9.81 ms^{-2} (in S.I. units). For simplicity, *g* may be given as 10 ms^{-2} in some problems.

The simplest form of projectile question is where all of the motion takes place in the vertical direction i.e. the problem is one-dimensional. Since the acceleration is taken to be constant, the one-dimensional forms of the constant acceleration formulae may be used.

In two-dimensional problems, such as when a projectile is fired at some angle α to the horizontal, the best approach is usually to **resolve** the velocity into **horizontal and vertical components**. In this way, the problem is divided into two one-dimensional problems. A consequence of the accepted model is that the horizontal component of velocity is taken to be constant.

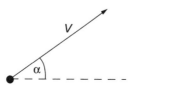

is represented by

The position of the **centre of mass** is given by

$$\bar{x} = \frac{\sum mx}{\sum m}, \text{ in vector form this becomes } \bar{\mathbf{r}} = \frac{\sum m\mathbf{r}}{\sum m}$$

In a uniform gravitational field, which is the standard model for A-level questions, the position of the centre of mass coincides with the **centre of gravity**, which is the point through which the weight of an object appears to act.

It should *not* be assumed, however, that the centre of mass of an object is at its *geometrical* centre, described as the **centroid**, unless the object is known to be uniform.

The **work done** W by a constant force F acting at an angle α to the direction of motion is given by $W = Fd \cos\alpha$, where d is the distance moved by the point of application of the force.

The S.I. unit of work is the joule and it should be noted that this is the same as the unit of energy. The effect of doing work on a body is to change its level of energy and, conversely, a body which possesses energy has the capacity to do work.

The **kinetic energy** (K.E.) of a body is the energy it has by virtue of its motion. K.E. $= \frac{1}{2}mv^2$.

The **potential energy** (P.E.) of a body is the energy it has by virtue of its position. For example, if an object of weight mg is raised, without resistance, to a height h above some fixed point then the work done is mgh. The object then has the potential to do an equal amount of work in returning to its original position. The P.E. of such a body at height h above a fixed point is therefore mgh. If, once released, the object is allowed to fall freely then the subsequent loss of P.E. is converted into K.E. The speed of the object on returning to its original level would then be given by $\frac{1}{2}mv^2 = mgh$. This interchange, between work and the two forms of mechanical energy, forms the basis of many A-level questions on this topic and needs careful consideration.

Power is the rate of doing work and is given by $\dfrac{dW}{dt} = Fv\cos\theta$. (In S.I. units $1\text{ W} = 1\text{ J s}^{-1}$.)

The **momentum** of a body is the product of its mass and its velocity i.e. momentum $= m\mathbf{v}$. In a collision, momentum is conserved even though energy is generally lost.

The **impulse** of a force is equal to the change in momentum which it produces and is given by $m\mathbf{v} - m\mathbf{u}$. If the force $\mathbf{F}$ is constant and acts for time t then $\mathbf{F}t = m\mathbf{v} - m\mathbf{u}$.

If an elastic string or spring is extended a distance x beyond its **natural length** l, then the resulting **tension** T is given by **Hooke's Law** which may be expressed in the form

$$T = \frac{\lambda x}{l}$$

where λ is a constant called the **modulus of elasticity**. Note that when $x = l$, $T = \lambda$ and so λ represents the force required to make a string or spring double its length which may, typically, be given in terms of mg. It follows that λ is measured in Newtons. The same equation gives the **thrust** in a spring which is compressed by a distance x inside its natural length.

The corresponding amount of work done to produce this extension, or compression, is stored as **elastic potential energy** E.P.E. and is given by

$$\text{work done} = \frac{\lambda x^2}{2l} = \text{E.P.E stored}$$

An object of mass m which moves with constant speed v can only maintain a **circular path** of radius r if acted upon by a force of magnitude $\dfrac{mv^2}{r}$ directed towards the centre of the circle.

It follows that the magnitude of the radial acceleration is $\dfrac{v^2}{r}$ which may be written as $\omega^2 r$ where ω is the angular velocity, measured in radians per second.

If you need to revise this subject more thoroughly, see the relevant topics in the *Letts* A level *Mathematics Study Guide*.

1 Find the time taken for a golf-ball to fall a distance of 2 m from rest. (3)

UCLES

2 A boy throws a stone which breaks a window 2 seconds later. The speed of projection is 20 ms^{-1} and the angle of projection is 60°. Assuming that the motion can be modelled with constant acceleration, find the horizontal and vertical components of the velocity of the stone just before the impact. (4)

UCLES

3

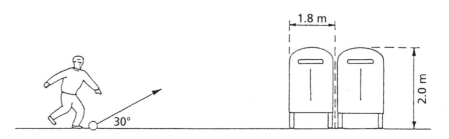

(In this question take $g = 10$ ms^{-2} and neglect air resistance.)

In an attempt to raise money for charity, participants are sponsored to kick a ball over some small vans. The vans are each 2 m high and 1.8 m wide and stand on horizontal ground.

One participant kicks the ball at an initial speed of 22 ms^{-1} inclined at 30° to the horizontal.

(i) What are the initial values of the vertical and horizontal components of velocity? (2)

(ii) Show that while in flight the vertical height y metres at time t seconds satisfies the equation $y = 11t - 5t^2$ and calculate at what times the ball is at least 2 m above the ground. (5)

The ball should pass over as many vans as possible.

(iii) Deduce that the ball should be placed about 3.8 m from the first van and find how many vans the ball will clear. (7)

Oxford & Cambridge (MEI, specimen)

4 A particle moves freely under gravity in the plane of a horizontal axis Ox and an upward vertical axis Oy. The particle is projected from O at time $t = 0$ with speed V and at an angle of elevation θ.

(a) (i) Write down expressions for the coordinates (x, y) of the particle at time t. (2)

(ii) Deduce that the equation of the path of the particle is

$$y = x \tan \theta - \left(\frac{gx^2}{2V^2} \right) \sec^2 \theta.$$ (2)

(b) A golf ball is projected from the point O with speed 40 ms^{-1} and lands at a point which is 10 metres below the level of O and at a horizontal distance of 120 metres from O. Taking $g = 10$ ms^{-2}, show that there are two possible angles of projection, one of which is $\tan^{-1}\left(\frac{1}{3}\right)$. Find the other. (6)

NEAB

5 A lift of mass 950 kg is carrying a woman of mass 50 kg.

(a) The lift is ascending at a uniform speed. Calculate:

(i) the tension in the lift cable; (3)

(ii) the vertical force exerted on the woman by the floor of the lift. (2)

(b) Sometime later the lift is ascending with a downward acceleration of 2 ms⁻². Calculate:

(i) the tension in the lift cable; (4)

(ii) the vertical force exerted on the woman by the floor of the lift. (3)

Oxford

6

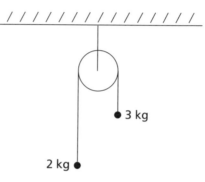

A light string passes over a pulley and has bodies of masses 2 kg and 3 kg attached to its ends. The system is released from rest in the position shown in the diagram. Find the tension in the string while the bodies are moving. (4)

State any two assumptions necessary for your method to be valid. (2)

UCLES

7

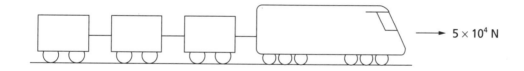

The diagram shows a goods train consisting of an engine of mass 40 tonnes and three trucks of 20 tonnes each. The engine is producing a driving force of 5×10^4 N, causing the train to accelerate. The ground is level and resistance forces may be neglected.

(i) Find the acceleration of the train. (4)

(ii) Draw a diagram to show the forces acting on the truck next to the engine. (3)

(iii) Find the tensions in each of the three couplings. (3)

The brakes on the middle truck are faulty and suddenly engage, causing a resistance of 10^4 N.

(iv) What effect does this have on the tension in the coupling to the last truck? (4)

Oxford & Cambridge (MEI, specimen)

8

The figure shows a footbridge across a stream, which consists of a plank *AB*, of length 5 m and mass 140 kg, supported at the ends *A* and *B* which are at the same level. When a man of mass 100 kg stands at the point *C* on the bridge, the magnitude of the force exerted by the support at *A* is twice the magnitude of the force exerted by the support at *B*.

(a) Suggest suitable models for the plank and the man in order to determine the magnitude of the force at *B* and the distance *AC*.

Determine estimates for

(b) the magnitude of the force exerted by the support at B,

(c) the distance of the man from A. (9)

ULEAC

9

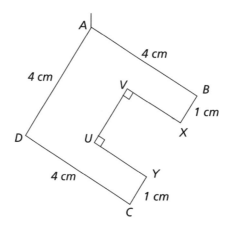

The diagram shows an ear-ring made from a uniform square lamina *ABCD*, which has side of length 4 cm. Points *X* and *Y* are on the side *BC* and such that *BX* = *CY* = 1 cm. The square portion *XYUV* is removed and the resulting ear-ring is suspended from the corner *A*. The ear-ring hangs in equilibrium.

The centre of mass of this ear-ring is *G*.

(a) State the distance, in cm, of *G* from *AB*.

(b) Find the distance, in cm, of *G* from *AD*.

(c) Find, to the nearest degree, the acute angle made by *AD* with the downward vertical. (11)

ULEAC

10 Two particles *P* and *Q*, each of mass *m*, are moving in the same direction along the same straight line with constant speeds $5u$ and u respectively. The particles collide and, after the collision, continue to move in the same direction as before, the speed of *Q* now being twice the speed of *P*.

(a) Show that the kinetic energy lost in the collision is $3mu^2$.

(b) Find the magnitude of the impulse exerted by *P* on *Q* in the collision. (10)

ULEAC

11 A light elastic rope of natural length a and modulus mg has one end fastened to a student, of mass m, and the other end to a fixed point A. The student stands at the edge of a platform, with the rope vertical, at a point O which is at a distance a below A. At a given instant the student steps off the platform and falls vertically. Show that when the rope has total length $a + x$, where $x > 0$, the speed of the student is given by v, where

$$av^2 = 2agx - gx^2.$$ (4)

Hence, or otherwise, find

(a) the maximum distance of the student below A; (2)

(b) the maximum speed of the student; (2)

(c) the maximum tension in the rope. (2)

AEB

12 A car has mass 720 kg and, at all speeds and on all roads, the resistance to motion of the car has magnitude 800 N.

(a) Find, in kW, the power output of the car when it is moving at a constant speed of 28 ms^{-1} on a level road.

With the same power output, the car now ascends a hill of fixed gradient at a constant speed of 20 ms^{-1}.

(b) Show that, to the nearest tenth of a degree, the angle of inclination of the hill to the horizontal is 2.6°. (9)

ULEAC

13 A toboggan of mass 15 kg carries a child of mass 25 kg. It starts from rest on a snow slope of inclination 10°. Given that the acceleration is 1.2 ms^{-2} and that air resistance may be ignored, find the coefficient of friction. (5)

Find the speed when it has moved 50 m from rest, and find the time taken. (4)

Having reached the bottom of the slope, the toboggan and child are pulled back up the slope, at a constant speed, by a light rope which is parallel to a line of greatest slope. Find the tension in the rope. (3)

UCLES

14 A particle P moves in a straight line in such a way that, at time t seconds, its velocity v ms^{-1} is given by

$$v = 12t - 3t^2, \quad 0 \le t \le 5,$$

$$v = -375t^{-2}, \quad t > 5.$$

When $t = 0$, P is at the point O.

Calculate the displacement of P from O

(a) when $t = 5$,

(b) when $t = 6$. (10)

ULEAC

15 The force **F** acting on a particle P at time t is given by $\mathbf{F} = 4t^2\mathbf{i} + (2t-3)\mathbf{j}$ where **i** and **j** are perpendicular unit vectors.

(i) If the particle has mass 0.2 kg find

    (a) its acceleration when $t = 2$ seconds, and

    (b) its velocity at time t if its initial velocity is $x\mathbf{i} + y\mathbf{j}$.   (7)

(ii) A second particle Q has position vector **r**, at time t, given by

$$\mathbf{r} = (t^2 + 2t)\mathbf{i} + (t^3 - 4t)\mathbf{j}.$$

Find its velocity at time t.   (3)

(iii) If the velocity of P relative to Q, at $t = 3$ seconds, is $180\mathbf{i} + 30\mathbf{j}$ find the values of x and y.   (4)

                                       NICCEA

16 A particle of mass 3 kg moves under the action of a single force **F** so that its position vector at time t is given by

$$\mathbf{r} = \cos 2t\mathbf{i} + \sin 2t\mathbf{j} - 6t\mathbf{k}.$$

(a) (i) Find the velocity and the momentum of P at time t.   (3)

    (ii) Express **F** in terms of t.   (2)

    (iii) Show that the direction of **F** is perpendicular to the velocity for all values of t.   (2)

(b) Briefly describe the motion of P given that the direction of **k** is vertically upwards.   (2)

                                       NEAB

17 A space-ship S, of mass M, is orbiting the moon 10^6 m above its surface with constant speed v ms^{-1}. In a preliminary model of this situation the moon is modelled as a sphere of radius 2×10^6 m, the space-ship as a particle and the acceleration due to gravity of the moon is modelled by the constant value 1.6 ms^{-2}. The space-ship travels round the moon in a circular orbit.

(a) Estimate, to 3 significant figures, the value of v.

A more refined model of the gravitational force **F**, experienced by S, is

$$\mathbf{F} = \frac{Mk}{r^2},$$

where **F** is directed towards O, the centre of the moon, and k is a constant. Given that the acceleration due to gravity at the moon's surface is of magnitude 1.6 ms^{-2},

(b) show that $k = 6.4 \times 10^{12}$ m^3 s^{-2}.

(c) Find a revised estimate for v.   (10)

                                       ULEAC

In order to understand a real situation, try to fit a **statistical model** to the data which will enable you to **make predictions** and **draw conclusions**. In deciding which model would be suitable, take account of the type of variable being considered (discrete or continuous) and the conditions under which it exists. Look at the model critically, and consider **refining** it in the light of practical experience.

The **discrete random variable** X denotes the **number of occurrences** of a particular event. The probabilities associated with these, written $p(x)$ or $P(X = x)$ are summarised in a **probability distribution**, a special property of which is that the sum of the probabilities, $\Sigma P(X = x)$ is 1.

The mean μ is known as the **expectation, $E(X)$**, where $E(X) = \Sigma x P(X = x)$.

The **variance, Var(X)**, is given by $\mathrm{Var}(X) = E\left(X^2\right) - \mu^2$, where $E\left(X^2\right) = \Sigma x^2 P(X = x)$.

Remember that the **standard deviation**, $\sigma = \sqrt{\mathrm{variance}}$.

The **cumulative distribution** function, $F\left(x_0\right) = P\left(X \le x_0\right)$, gives a running total of probabilities.

The probability function for the **continuous random variable X**, where $a < x < b$, is usually written $f(x)$ and is represented by a curve. Probabilities are given by the **area under the curve** where $P(c < X < \mathrm{d}) = \int_c^d f(x)\mathrm{d}x$ and the total area under the curve is 1.

$$E(X) = \int_a^b x f(x)\mathrm{d}x \text{ and } \mathrm{Var}(X) = E\left(X^2\right) - \mu^2, \text{ where } E\left(X^2\right) = \int_a^b x^2 f(x)\mathrm{d}x.$$

The **cumulative distribution function** is given by $F\left(x_0\right) = P\left(X \le x_0\right) = \int_a^{x_0} f(x)\mathrm{d}x$, and $F(m) = 0.5$ where m is the median.

To obtain the probability function from the cumulative function, use $f(x) = F'(x)$.

It is possible to find the **expectation and variance of simple functions of the variable:**

$$E(aX \pm b) = aE(X) \pm b \qquad \mathrm{Var}(aX \pm b) = a^2 \mathrm{Var}(X),$$

and **linear combinations of two random variables :**

$$E(aX \pm bY) = aE(X) \pm bE(Y) \qquad \mathrm{Var}(aX \pm bY) = a^2 \mathrm{Var}(X) + b^2 \mathrm{Var}(Y).$$

However, remember that the variance result holds only if X and Y are *independent.*

The **binomial distribution B(n, p)** is used to model the number of successes in n independent trials when the probability of success, p, is constant.

Probabilities are given by $P(X = x) = \dbinom{n}{x} q^{n-x} p^x$, with $x = 0, 1, 2, ..., n$.

$E(X) = np$ and $\mathrm{Var}(X) = npq$, where $q = 1 - p$.

The **geometric distribution Geo(p)** is used to model the number of trials of an experiment up to and including the first success, where p is the probability of success.

Probabilities are given by $P(X = x) = q^{x-1}p$, with $x = 1, 2, 3, ...$

$E(X) = \dfrac{1}{p}$ and $\mathrm{Var}(X) = \dfrac{q}{p^2}$.

The **Poisson distribution Po(μ)** is used to model the distribution of random events, where μ is the average number of successes. Probabilities are given by $P(X = x) = e^{-\mu} \dfrac{\mu^x}{x!}$, where $x = 0, 1, 2, 3, \ldots$, $E(X) = \mu$ *and* $\text{Var}(X) = \mu$.

The sum of independent Poisson variables is also a Poisson variable.

The Poisson distribution can also be used as an **approximation to the binomial distribution**, where $X \sim \text{Po}(np)$. This applies when n is large (say > 50) and p is small (say < 0.1) such that np is approximately equal to npq.

The **normal distribution** is the most important **continuous distribution** and it is used to model many situations, particularly in the natural and physical world.

We write $X \sim N(\mu, \sigma^2)$, where $E(X) = \mu$ and $\text{Var}(X) = \sigma^2$. To find probabilities, relate X to the **standard normal variable Z**, where $Z = \dfrac{X - \mu}{\sigma}$, and use tables. There are several different formats of the tables, so you must make sure that you are familiar with the one that you will be given in the examination.

The normal distribution can be used as an **approximation**
(a) **to the binomial distribution**: $X \sim N(np, npq)$ when n is large (say > 30) and p is not too far from 0.5. Note that the approximation is better, the nearer that p is to 0.5.
(b) **to the Poisson distribution**: $X \sim N(\mu, \mu)$ when μ is large, say > 20.

In both cases, it is necessary to use a **continuity correction**, since a continuous distribution is being used as an approximation to a discrete one.

An important property of the normal variable is that a linear combination of independent normal variables is also normal. This enables us to form the **distribution of the sample mean, $\bar{X}$,** where $\bar{X} \sim N(\mu, \sigma^2/n)$. By the **Central Limit Theorem**, this result also holds when the parent distribution is not normally distributed, provided that n, the size of the sample, is large.

When the population variance σ^2 is not known, it can be **estimated** by multiplying the sample variance by $n/(n-1)$. Note that an unbiased estimate for the population mean μ is the sample mean, $\bar{x}$.

We can also form **confidence intervals**, for example, a 95% confidence interval for μ, based on the sample mean, is $\bar{x} \pm 1.96 \dfrac{\sigma}{\sqrt{n}}$. Note that the probability that the interval includes μ is 95%.

When performing a **significance test** there is a standard procedure which it is helpful to follow: First decide on the **null hypothesis H_0** and the **alternative hypothesis H_1** (taking account of whether the test is **one-tailed or two-tailed**), and also α, **the level of significance** of the test. Now state the distribution of the variable, according to the null hypothesis, and decide on the **criterion for rejecting H_0**. In general, the null hypothesis is rejected if it lies in the 'tail end' of the distribution and the **critical value** to denote this region is determined by the level of the test. Finally the sample value is tested, and a conclusion made.

When performing a test, a **type 1** error occurs when we *wrongly reject* H_0 (this is given by the level of the test) and a **type 2** error occurs when we *wrongly accept* H_0. To calculate this, a definite value for H_1 must be stated.

The χ^2 **test** (chi-squared) is used to test how well a theoretical distribution, such as a uniform, binomial, Poisson or normal, models a given practical situation (sometimes known as a **goodness of fit** test), and to test for independence in a **contingency table**.

The null hypothesis H_0 is made and the difference between the observed values (O) and those expected according to the null hypothesis (E) are investigated by calculating the following approximation for the χ^2 statistic: $\sum \dfrac{(O-E)^2}{E}$. The null hypothesis is rejected if this value lies in the **critical region**, where the **critical value** is found from tables and depends on the level of the test and also the number of **degrees of freedom,** υ = number of cells – number of restrictions. An important rule to remember is that all expected frequencies must be at least 5. If they are not, then cells must be combined. You need to consider this *before* υ is calculated.

If $\upsilon = 1$ (as in a 2×2 contingency table), Yates' correction should be used, in which case the test statistic is $\sum \dfrac{(|O-E|-0.5)^2}{E}$.

Bivariate data can be represented on a **scatter diagram** and lines of best fit drawn. The most usual are the **least squares regression lines of y on x** and **of x on y**. Remember that both lines go through $(\bar{x}, \bar{y})$. The regression line of y on x can be found by using

❶ the **regression coefficient**, where $y - \bar{y} = \dfrac{s_{xy}}{s_x^2}(x - \bar{x})$, or

❷ the **normal equations**: $\Sigma y = na + b\Sigma x$, $\Sigma xy = a\Sigma x + b\Sigma x^2$.

You should be familiar with the format provided in your reference booklet. You may have to use raw data or summarised statistics, but it is also useful to be proficient in calculating in LR mode on your calculator. The appropriate regression line can be used to estimate a value, but you should be wary of estimating values which are *outside the range* of the given data. Note also that **outliers** can influence the equation.

An idea of **linear correlation** can be formed from the scatter diagram, ranging from perfect negative correlation (all the points lying on a straight line with negative gradient), to no correlation (completely random scatter), to perfect positive correlation (all the points lying on a straight line with positive gradient).

The **product-moment correlation coefficient, r,** can be calculated

using $r = \dfrac{s_{xy}}{s_x s_y}$, where $-1 \le r \le 1$.

Another useful coefficient is **Spearman's coefficient of rank correlation, r_s.** Both x and y data are ranked in order and for each of the n pairs of values, d is calculated, where d = rank of x – rank of y. The coefficient is given by $r_s = \dfrac{6\Sigma d^2}{n(n^2 - 1)}$.

This measure is particularly useful when only the ranks of x and y are known, rather than their actual values.

If you need to revise this subject more thoroughly, see the relevant topics in the *Letts* A level Mathematics Study Guide.

1 A commuter, who caught the same train to work every day, kept a record over six months and found that the train was late on 49 days out of a total of 126 working days. Use a binomial probability model to estimate the probability that, in a five day working week, the train will be late exactly three times. (4)

State briefly, giving reasons, whether a binomial distribution is likely to be a satisfactory or unsatisfactory mathematical model in this situation. (2)

UCLES

2 During weekends, arrivals of patients at the casualty department of a hospital occur at random at a rate of 5 per hour.

(a) Find the probability that during any one hour period in a weekend exactly 3 patients will arrive.

A patient arrives at exactly 10.00 am.

(b) Find the probability that the next patient arrives before 10.15 am. (6)

ULEAC

3 The discrete random variable X takes integer values 1, 2, 3, 4, 5, 6, 7.
$$P(X=1)=\frac{1}{13}, \; P(X=2)=\frac{3}{13}$$
$$P(X=3)=P(X=4)=P(X=6)=P(X=7)=2\,P(X=5)$$

(a) Calculate the mean and variance of X. (5)

(b) Deduce the mean and variance of $X + 3$. (2)

NEAB

4 (a) The heights of foxglove plants growing in **woodland** are known to be normally distributed with mean 27.5 cm and standard deviation 3.5 cm. Calculate the probability that a randomly chosen foxglove is more than 35 cm high. (4)

(b) On **riverbanks**, the heights of foxgloves are distributed normally with mean 32.0 cm and standard deviation 4 cm. Calculate the probability that a randomly chosen riverbank foxglove is less than 35 cm high. (2)

(c) Calculate the probability that a randomly selected woodland foxglove is taller than a randomly selected riverbank foxglove. (6)

Oxford

5 (a) Customers enter a car showroom at a roughly constant rate and each has a small probability of buying a car, independently of sales to other customers. Suggest a distribution for the number of sales made in a week. (1)

(b) The numbers of sales in a week in two car showrooms have means 2.4 and 3.6, respectively.

(i) Calculate the probability that the first showroom will sell 3 cars and the second 5 cars in a week. State any assumption you make. (4)

(ii) Find the probability that the two showrooms will sell a total of fewer than 8 cars in a given week. (2)

(iii) Determine an approximate value for the probability that the first showroom will sell at least 30 cars in a period of 10 weeks. (4)

NEAB

6 In an attempt to economise on her telephone bill, Debbie times her calls and ensures that they never last longer than 4 minutes.

The length of the calls, T, minutes, may be regarded as a random variable with probability density function

$$f(t) = \begin{cases} kt & 0 < t \le 4 \\ 0 & \text{otherwise} \end{cases}$$

where k is a constant.

(a) (i) Show that $k = 0.125$. (1)

    (ii) Find the mean and standard deviation of T. (5)

    (iii) Find the probability that a call lasts between 3 and 4 minutes. (1)

    (iv) What is the probability that, of 5 independent calls, exactly 3 last between 3 and 4 minutes? (3)

Calls are charged at a rate of 6p per call plus 4p for each complete minute that the call lasts.

(b) (i) Copy and complete the following table

length of call, mins	probability	cost, pence
0–1		
1–2		
2–3		
3–4		18

(3)

    (ii) Find the mean and standard deviation of the cost, in pence, of a call. (4)

AEB

7 Among the blood cells of a certain animal species, the proportion of cells which are of type A is 0.37 and the proportion of cells which are of type B is 0.004.

(a) Find the probability that in a random sample of 8 blood cells, at least 2 will be of type A.

Using suitable approximations find the probability that

(b) in a random sample of 200 blood cells the total number of type A and type B cells is at least 81,

(c) in a random sample of 300 blood cells there will be at least 4 cells of type B. (17)

ULEAC

8 An enthusiastic gardener claimed that she could never work in the garden at the weekend because 'It always rains on Saturday and Sunday when I'm at home and it's always fine on weekdays when I'm not!'. She noted the weather for the next month and recorded that, out of 10 wet days, 5 were either a Saturday or a Sunday. The gardener's claim may be modelled by regarding her observation as a single sample from a B(10, p) distribution. Given that one would expect two out of every seven wet days to be either a Saturday or a Sunday, the null hypothesis, $p = \frac{2}{7}$, may be tested against the alternative hypothesis, $p > \frac{2}{7}$. Carry out a hypothesis test to test her claim at the 10% significance level. (6)

UCLES

9 A machine is supposed to produce keys to a nominal length of 5.00 cm. A random sample of 50 keys produced by the machine was such that $\Sigma x = 250.50$ cm and $\Sigma x^2 = 1255.0290$ cm^2, where x denotes the length of a randomly chosen key produced by the machine.

(i) Calculate unbiased estimates of the mean and variance of the length of keys produced by the machine, giving your answers correct to an appropriate degree of accuracy. (3)

(ii) Calculate a 95% symmetric confidence interval for the mean length of keys produced by the machine, giving your answers correct to an appropriate degree of accuracy. (5)

(iii) Carry out a hypothesis test at the 5% significance level to test the assertion that the machine is producing keys which are too long, giving suitable null and alternative hypotheses and stating your conclusion clearly. (5)

UCLES

10 The following table shows the number of girls in families of 4 children:

Number of girls	0	1	2	3	4
Frequency	15	68	69	38	10

A researcher suggests that a binomial distribution with $n = 4$ and $p = 0.5$ could be a suitable model for the number of girls in a family of 4 children.

(a) Test the researcher's suggestion at the 5% level, stating your null and alternative hypotheses clearly.

The researcher decides to progress to a more refined model and retains the idea of a binomial distribution but does not specify the value of p, the probability that the child is a girl.

(b) Use the data in the table to estimate p.

The researcher used the value of p in (b), and the refined model, to obtain expected frequencies and found $\sum \frac{(O-E)^2}{E} = 2.47$. (There was no pooling of classes.)

(c) Test, at the 5% level, whether the binomial distribution is a suitable model of the number of girls in a family of 4 children.

(d) A family planning clinic has a large number of enquiries from families with 3 boys who would like a fourth child in the hope of having a girl but they believe their chances are very small. What advice can the researcher give on the basis of the above tests? (18)

ULEAC

11 Delegates who travelled by car to a conference were asked to report the distance, d miles, they travelled and the time, t minutes, taken. A random sample of the values reported are given in the table below.

Distance d	113	14	98	130	75	120	143	55	127
Time t	130	25	180	148	100	120	196	48	165

$[\Sigma d = 875, \Sigma t = 1112, \Sigma d^2 = 99\ 097, \Sigma t^2 = 164\ 174, \Sigma dt = 125\ 443.]$

(i) Plot a scatter diagram of the data, with d on the horizontal axis and t on the vertical axis. (4)

(ii) Calculate $\bar{d}$, the mean of d, $\bar{t}$, the mean of t and the regression line of time on distance, giving your answer in the form $t = a + bd$, where a and b are constants to be determined. Give your answers correct to three significant figures.
Plot the point $(\bar{d}, \bar{t})$ and the regression line on your scatter diagram. (6)

(iii) Use the linear model obtained in (ii) to estimate the average time taken by a delegate who travels 100 miles by car. (2)

(iv) Calculate the product moment correlation coefficient for the given data and interpret the result of this calculation in terms of your scatter diagram. (4)

UCLES

Answers

1 ALGEBRA

Answer	Mark	Examiner's tip

1 (i) $a^k = \sqrt[3]{(a)^4} \div a$

Apply the laws of indices to express the RHS of the equation in the same form as the LHS.

$a^k = a^{\frac{4}{3}} \div a^1$ — **2**

$a^k = a^{\frac{1}{3}}$

The indices may now be compared directly.

$\underline{k = \frac{1}{3}}$ — **2**

(ii) $27^x = 9^{(x-1)}$

In essence, this is similar to part (i) but, this time, it is helpful to re-write both sides using the same number as base.

$(3^3)^x = (3^2)^{(x-1)}$ — **2**

$3x = 2(x-1)$

$\underline{x = -2}$ — **2**

2 (a) $2x^{\frac{1}{3}} = x^{-\frac{2}{3}}$

The rational index is at the heart of this problem. A good first move is to multiply both sides of the equation

$2x = 1$ — **3**

$\underline{x = \frac{1}{2}}$ — **1**

by $x^{\frac{2}{3}}$ which has the effect of clearing the fractions.

(b) $3^y = 6$

Taking logs of both sides converts the equation to linear form.

$y\ln 3 = \ln 6$ — **2**

$\underline{y = 1.63}$ (correct to 3 s.f.) — **3**

State the answer to the required degree of accuracy.

3 $y = a(b^x)$

$\ln y = \ln(ab^x)$

$\ln y = \ln a + x\ln b$ — **1**

The hint contained in the question is that we should take the natural logarithm of both sides to convert the equation to gradient-intercept form.

The gradient is given by $\ln b = 1.8$

$\Rightarrow \underline{b = e^{1.8} = 6.05}$ (to 3 s.f.) — **2**

and the y-intercept is given by

$\ln a = 4.1$

$\Rightarrow \underline{a = e^{4.1} = 60.3}$ (to 3 s.f.) — **2**

4 (i) (A) The population *increases* each year — **1**
(B) The population *doesn't change* — **1**
(C) The population *reduces* each year. — **1**

(ii)

Year	1910	1920	1930	1940	1950	1960
n	10	20	30	40	50	60
$\log_{10}P_n$	1.57	1.64	1.72	1.81	1.89	1.97

2

Answer	Mark	Examiner's tip

(iii) For both marks you would have to plot all of the points accurately on the axes provided.

2

(iv) Your answer should make reference to the fact that the graph is a straight line.

2

Care should be taken to draw an accurate graph, because the readings taken from it later will be needed in calculations to check the model, $\log_{10} P_n = \log_{10} P_0 + n\log_{10} k$ which is in linear form, and so the model predicts that the graph of $\log_{10} P_n$ against n should be linear.

(v) From the graph, the y-intercept is given by $\log_{10} P_0 = 1.49 \Rightarrow \underline{P_0 = 31}$
(million, to nearest million)

and $\log_{10} k = 0.008 \Rightarrow \underline{k = 1.0186}$
(correct to 5 s.f.)

3

Any greater accuracy couldn't be justified, based on the accuracy of the given data.

This is a high degree of accuracy, but it fits the given data very well.

(vi) Using the figures for P_0 and k from (v) in the model, there is close agreement with the actual figures. However, the predictions for 1970, 1980 and 1990 are far too high, which suggests that the pattern of population growth has changed.

2

There is no need to go into too much detail - be guided by the number of marks available.

5 Let $f(x) = x^3 + ax^2 + bx - 4$

then $f(2) = 0 \Rightarrow 8 + 4a + 2b - 4 = 0$

$\Rightarrow 4a + 2b + 4 = 0$

$\Rightarrow 2a + b + 2 = 0$ (1)   **1**

and $f(-2) = 0 \Rightarrow -8 + 4a - 2b - 4 = 0$

$\Rightarrow 2a - b - 6 = 0$ (2)   **1**

(1) + (2) gives $4a - 4 = 0 \Rightarrow \underline{a = 1}$ **1**

Substituting for a in (1) gives $\underline{b = -4}$ **1**

Substituting for a and b in the original equation gives $x^3 + x^2 - 4x - 4 = (x - 2)(x + 2)g(x)$ (where $g(x)$ is linear).

i.e. $x^3 + x^2 - 4x - 4 = (x^2 - 4)g(x)$

and so $g(x) = \underline{x + 1}$ is the required linear factor.

2

Consider introducing function notation as a means of making your method clear.

Check that $f(-1) = 0$.

6 (a) $f(-2) = 2(-2)^3 - 3(-2)^2 - 11(-2) + 6$

$= -16 - 12 + 22 + 6 \doteq 0.$

∴ By the factor theorem $(x + 2)$ is a factor of $f(x)$.

1

It's important to show some working – don't simply state that $f(-2) = 0$.
Make a point of referring to the factor theorem.

(b) $f(x) = (x + 2)(2x^2 - 7x + 3)$

$\underline{= (x + 2)(2x - 1)(x - 3)}$

2

Answer	Mark	Examiner's tip

(c)

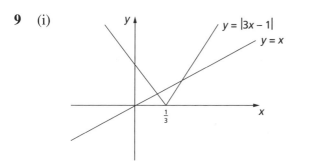

From the graph,
$f(x) \leq 0 \Rightarrow x \leq -2$ or $\frac{1}{2} \leq x \leq 3.$

2

The inequality, $f(x) \leq 0$, is solved by considering where the graph of the function lies on or below the *x*-axis.

There are only two marks available here so just a quick sketch showing the general shape and the points of intersection with the *x*-axis will suffice.

7 $x^2 - 4x + 9 = (x-2)^2 - 4 + 9$
$= (x-2)^2 + 5$

2

Minimum value of $x^2 - 4x + 9$ is 5 and so the maximum value of $f(x)$ is $\frac{1}{5}.$

1

This result can be checked by using a graphics calculator to draw the graph of $y = f(x)$ and locating the maximum value with the trace function.

8 Let $f(x) = 2x^3 + ax^2 + 16x + 6$

then $f(x) = (2x + 1)g(x)$ where $g(x)$ is a polynomial function.

It follows that $f(-\frac{1}{2}) = 0$

i.e. $2(-\frac{1}{2})^3 + a(-\frac{1}{2})^2 + 16(-\frac{1}{2}) + 6 = 0$

1

$\Rightarrow -\frac{1}{4} + \frac{a}{4} - 8 + 6 = 0 \Rightarrow -1 + a - 8 = 0$
$\Rightarrow \underline{a = 9}$

1

$f(x) = (2x + 1)(x^2 + 4x + 6)$

i.e. the required quadratic factor is $x^2 + 4x + 6.$

Now, $x^2 + 4x + 6 = (x+2)^2 - 4 + 6$
$= (x+2)^2 + 2$

2

which is positive for all real values of x (the minimum value is 2).

2

The introduction of function notation can, again, make the task of filling in the necessary steps, to justify the end result, easier.

Having completed the square, some justification for concluding that the quadratic is always positive should be given. One way is to interpret the result in terms of the minimum value.

9 (i)

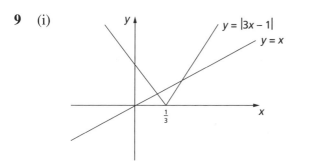

3

The graph of $y = |3x - 1|$ is the same as the graph of $y = 3x - 1$ wherever $3x - 1 \geq 0$, and corresponds to the graph of $y = -(3x - 1)$ when $3x - 1 < 0.$

(ii) The larger value of *x* corresponds to the solution of $3x - 1 = x$ (since at this point $3x - 1 > 0$).

$\Rightarrow x = \frac{1}{2}$

It is important to recognise how the modulus function behaves in order to produce **two** equations to be solved.

Answer	Mark	Examiner's tip

The smaller value of x corresponds to the solution of $-(3x - 1) = x$
(since at this point $3x - 1 < 0$).

$$\Rightarrow x = \tfrac{1}{4}$$

3

(iii) $|3x - 1| > x \Rightarrow x < \tfrac{1}{4}$ or $x > \tfrac{1}{2}$ **3**

The solutions occur where the graph of $y = |3x - 1|$ lies above the graph of $y = x$.

10 (a)

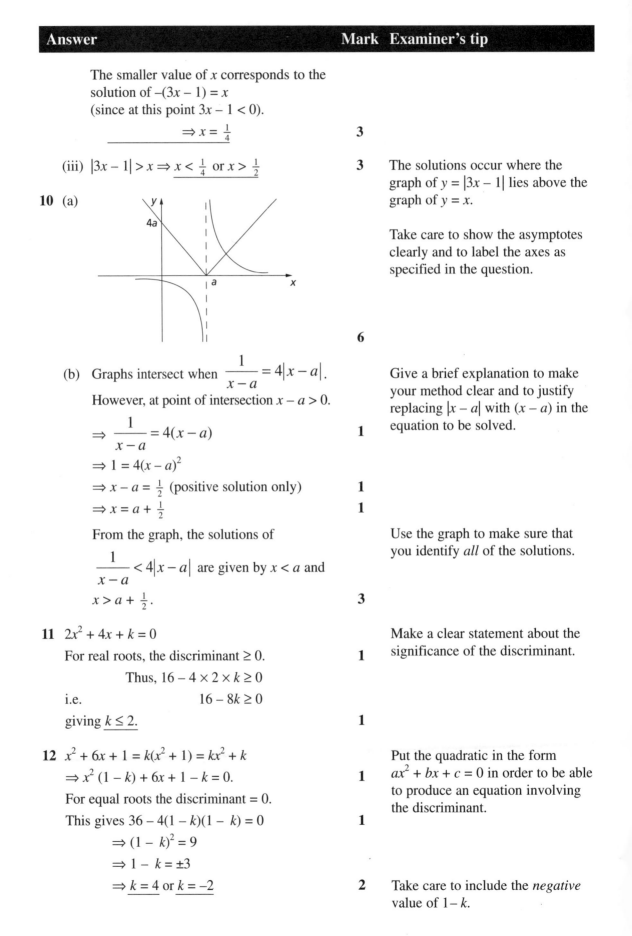

6

Take care to show the asymptotes clearly and to label the axes as specified in the question.

(b) Graphs intersect when $\dfrac{1}{x - a} = 4|x - a|$.

However, at point of intersection $x - a > 0$.

$$\Rightarrow \frac{1}{x - a} = 4(x - a)$$ **1**

$$\Rightarrow 1 = 4(x - a)^2$$

$$\Rightarrow x - a = \tfrac{1}{2} \text{ (positive solution only)}$$ **1**

$$\Rightarrow x = a + \tfrac{1}{2}$$ **1**

Give a brief explanation to make your method clear and to justify replacing $|x - a|$ with $(x - a)$ in the equation to be solved.

From the graph, the solutions of

$$\frac{1}{x - a} < 4|x - a| \text{ are given by } x < a \text{ and}$$

$$x > a + \tfrac{1}{2}.$$ **3**

Use the graph to make sure that you identify *all* of the solutions.

11 $2x^2 + 4x + k = 0$

For real roots, the discriminant ≥ 0.   **1**

Thus, $16 - 4 \times 2 \times k \geq 0$

i.e. $16 - 8k \geq 0$

giving $k \leq 2$.   **1**

Make a clear statement about the significance of the discriminant.

12 $x^2 + 6x + 1 = k(x^2 + 1) = kx^2 + k$

$$\Rightarrow x^2(1 - k) + 6x + 1 - k = 0.$$ **1**

For equal roots the discriminant $= 0$.

This gives $36 - 4(1 - k)(1 - k) = 0$ **1**

$$\Rightarrow (1 - k)^2 = 9$$

$$\Rightarrow 1 - k = \pm 3$$

$$\Rightarrow k = 4 \text{ or } k = -2$$ **2**

Put the quadratic in the form $ax^2 + bx + c = 0$ in order to be able to produce an equation involving the discriminant.

Take care to include the *negative* value of $1 - k$.

Answer	Mark	Examiner's tip

13 (a) $y = x^2 + px + q$

Substituting $x = 2$, $y = 2$ gives

$2 = 4 + 2p + q$

which simplifies to $2p + q + 2 = 0$.   (1)   **1**

Replacing x and y in the equation of the parabola, with the coordinate values, produces the required result.

(b) $\dfrac{dy}{dx} = 2x + p$ **1**

When $x = 2$, $\dfrac{dy}{dx} = 1$ **1**

$\Rightarrow 4 + p = 1 \Rightarrow \underline{p = -3}$ **1**

From (1) $-6 + q + 2 = 0 \Rightarrow \underline{q = 4}$.   **1**

Hence, the equation of the parabola is

$\underline{y = x^2 - 3x + 4}$ **2**

The gradient of $y = x$ is 1 and so $\dfrac{dy}{dx} = 1$ at the point (2, 2).

(c) $b^2 - 4ac = p^2 - 4q = 9 - 16 = \underline{-7}$ (< 0).   **1**

The negative value of the discriminant confirms that the parabola doesn't meet the x-axis since the equation has no real roots.   **1**

14 $x + y = 2$ (1)

$x^2 + 2y^2 = 11$ (2)

from (1), $x = 2 - y$ **1**

$\Rightarrow x^2 = 4 - 4y + y^2$ (3)   **1**

Substituting for x^2 in (2) gives

$4 - 4y + y^2 + 2y^2 = 11$ **1**

$\Rightarrow 3y^2 - 4y - 7 = 0$

$\Rightarrow (3y - 7)(y + 1) = 0$ **1**

$\Rightarrow y = \frac{7}{3}$ or $y = -1$.

When $\underline{y = \frac{7}{3}, x = -\frac{1}{3}}$ **1**

and when $\underline{y = -1, x = 3}$.   **1**

Substitute from the linear equation into the quadratic.

The idea is to produce a quadratic in one unknown.

When stating the solution, the corresponding values of x and y should be given in pairs.

15 (a) The points of intersection are given by the simultaneous solution of

$y = x$ (1)

and $x^2 + y^2 - 6x - 2y - 24 = 0$.   (2)

Substituting for y in (2) gives

$x^2 + x^2 - 6x - 2x - 24 = 0$ **1**

$\Rightarrow 2x^2 - 8x - 24 = 0$

$\Rightarrow x^2 - 4x - 12 = 0$

$\Rightarrow (x + 2)(x - 6) = 0$

$\Rightarrow x = -2$ or $x = 6$ **1**

There is a common factor of 2 in each term that can be cancelled to simplify the equation.

Answer	Mark	Examiner's tip

The coordinates of the points of intersection are

$A(6, 6)$ and $B(-2, -2)$.   **1**

Give the answer in the form specified in the question. It is clear from the given diagram which point is A and which is B.

(b) The centre of the circle, with AB as diameter, has coordinates $(2, 2)$.

It is useful to draw a diagram in order to find the centre and radius of the new circle. Note, the $(\text{radius})^2$ is given by Pythagoras.

Equation of circle is

$(x - 2)^2 + (y - 2)^2 = 32$ **3**

16 $\log_e (x + y) = 0 \Rightarrow x + y = 1$ (1)   **2**

$2 \log_e x = \log_e (y - 1)$ (2)

$\Rightarrow \log_e x^2 = \log_e (y - 1)$

$\Rightarrow x^2 = y - 1$ (3)   **3**

Substituting for y in (3) gives

$\quad x^2 = -x \Rightarrow x(x + 1) = 0$ **1**

$\quad\quad \Rightarrow x = 0$ or $x = -1$ (but $x \neq 0$ from (2))   **2**

The only solution is $x = -1$ and $y = 2$.   **2**

The complication here is that the simultaneous equations are both expressed in terms of logs. Having realised this, the first move is to use the basic properties of logs to re-write the equations in a more familiar form.

Note

$\log_e(1) = 0$

17

(a) Let

$$\frac{1 - x - x^2}{(1 - 2x)(1 - x)^2} \equiv \frac{A}{(1 - 2x)} + \frac{B}{(1 - x)^2} + \frac{C}{(1 - x)}$$

then

$1 - x - x^2 \equiv A(1 - x)^2 + B(1 - 2x) + C(1 - x)(1 - 2x)$

This relationship holds for all values of x.

Setting $x = 1$ gives $-1 = -B \Rightarrow B = 1$ **1**

$\quad x = \frac{1}{2}$ gives $\frac{1}{4} = \frac{1}{4} A \Rightarrow A = 1$ **1**

$\quad x = 0$ gives $1 = A + B + C \Rightarrow C = -1$.   **1**

Select x values that will produce the simplest equations in A, B and C.

Hence,

$$\frac{1 - x - x^2}{(1 - 2x)(1 - x)^2} \equiv \frac{1}{(1 - 2x)} + \frac{1}{(1 - x)^2} - \frac{1}{(1 - x)}$$ **1**

Complete this part of the question by substituting for A, B and C.

(b) $\dfrac{1}{(1 - 2x)} = 1 + 2x + 4x^2 + 8x^3 + \ldots$ **1**

provided $|2x| < 1$

$\quad$ i.e. $|x| < \frac{1}{2}$

These results are based on the binomial expansion of $(1 - x)^{-1}$ which is an important standard result.

Answer	Mark	Examiner's tip

$$\frac{1}{(1-x)} = 1 + x + x^2 + x^3 + \dots$$

provided $|x| < 1$ **1**

$$\frac{1}{(1-x)^2} = 1 + 2x + 3x^2 + 4x^3 + \dots$$

provided $|x| < 1$ **1**

Hence

$$\frac{1 - x - x^2}{(1-2x)(1-x)^2} = 1 + 3x + 6x^2 + 11x^3 + \dots$$ **2**

(c) The expansion is valid provided $|x| < \frac{1}{2}$.   **1**

18 (a) (i) $f(x) = 2x^3 - x^2 - 7x + 6$ **1**
$\Rightarrow f(1) = 2 - 1 - 7 + 6 = 0$
$2x^3 - x^2 - 7x + 6 = (x-1)(2x^2 + x - 6)$ **1**
$ = (x-1)(2x-3)(x+2)$ **2**

(ii)

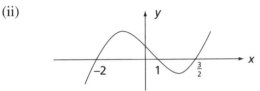

From the graph, $f(x) > 0$ when $-2 < x < 1$ **3**
and when $x > \frac{3}{2}$.

(b) $$\frac{x^2 + 2x + 7}{(2x+3)(x^2+4)} \equiv \frac{A}{(2x+3)} + \frac{Bx+C}{(x^2+4)}$$

$$\Rightarrow x^2 + 2x + 7 \equiv A(x^2 + 4) + (Bx + C)(2x + 3)$$

Setting up equations and solving in the
usual way gives $A = 1$, $B = 0$ and $C = 1$.   **3**

Hence

$$\int \frac{x^2 + 2x + 7}{(2x+3)(x^2+4)} dx = \int \frac{1}{(2x+3)} + \frac{1}{(x^2+4)} dx$$

$$= \tfrac{1}{2}\ln|2x+3| + \tfrac{1}{2}\arctan\frac{x}{2} + c$$ **3**

Examiner's tip column:

Note: $\dfrac{1}{(1-x)^2}$ is the derivative of $\dfrac{1}{(1-x)}$ and the corresponding series expansion can be found by differentiating $1 + x + x^2 + x^3 + \dots$ term by term.

The hint is that you should now use the factor theorem to identify $(x-1)$ as a factor of $f(x)$.

A sketch graph is helpful here. Note, the graph has the general shape of a cubic and crosses the x-axis when $x = 1, \frac{3}{2}, -2$ corresponding to the factors found in part (i).

Note: equations in A, B and C can be obtained either by substituting values for x or by comparing like terms. The aim, as always, is to produce the simplest equations to work with.

See the chapter on **integration** for advice on the techniques needed here.

Answer	Mark	Examiner's tip

19 (a)

$$f(x) \equiv \frac{x^2 + 6x + 7}{(x+2)(x+3)} \equiv \frac{(x^2 + 5x + 6) + (x+1)}{x^2 + 5x + 6}$$

1 — In this case, the numerator and denominator are both of degree 2 and so it is necessary to divide before attempting to use partial fractions.

$$\equiv 1 + \frac{x+1}{(x+2)(x+3)} \equiv A + \frac{B}{(x+2)} + \frac{C}{(x+3)}$$

where $A = 1$, $B = -1$, $C = 2$

3 — Since the denominators are linear, the values of B and C can be found quickly using the **cover-up method**.

(b) $\displaystyle \int_0^2 f(x)\,dx = \int_0^2 1 - \frac{1}{(x+2)} + \frac{2}{(x+3)}\,dx$

$$= \left[x - \ln|x+2| + 2\ln|x+3| \right]_0^2$$

2

$$= 2 - \ln 4 + 2\ln 5 - (0 - \ln 2 + 2\ln 3)$$

2

$$= 2 + \ln\left(\frac{5^2 \times 2}{4 \times 3^2} \right) = 2 + \ln\left(\frac{25}{18} \right)$$

3 — Note the need to use the log laws to present the final answer in the required form.

2 SEQUENCES AND SERIES

Answer	Mark	Examiner's tip

1 (i) The general term is given by
$u_k = a + (k-1)d$
$\Rightarrow 36 = a + 9d$ (1)

It's a good idea to label equations that will be needed later.

The sum of the first n terms is given by

$$S_n = \frac{n}{2}(a + l)$$

$$\Rightarrow 180 = \frac{10}{2}(a + a + 9d)$$

2

$$\Rightarrow 36 = 2a + 9d \quad (2)$$

From (1) and (2) $a = 0$ and $d = 4$

2 — Solve equations (1) and (2) simultaneously.

i.e. the first term is 0 and the common difference is 4.

(ii) $\displaystyle \sum_{r=1}^{1000} (3r - 1) = 2 + 5 + 8 + \ldots + 2999$

2 — Substitute $r = 1, 2, 3$ and 1000 to get an idea of how the series develops.

This is an A.P. with $a = 2$, $d = 3$, $n = 1000$ and $l = 2999$.

The fact that the series is an A.P. is very significant and worth stating.

$$S_{1000} = \frac{1000}{2}(2 + 2999)$$

$$= 1\,500\,500$$

1

Answer	Mark	Examiner's tip

2 (a) $S_\infty = \dfrac{a}{1-r}$ $S_8 = \dfrac{a(1-r^8)}{1-r}$

Express the given information using the standard notation and formulas to produce an equation.

$S_8 = \frac{1}{2}S_\infty \Rightarrow 1-r^8 = \frac{1}{2} \Rightarrow r^8 = \frac{1}{2}$

$\Rightarrow \underline{r = 0.917}$ (correct to 3 d.p.)   **3**

17 th term $= ar^{16} = 10$

Note: $r^{16} = (r^8)^2 = \left(\frac{1}{2}\right)^2 = \frac{1}{4}$.

$\Rightarrow a = \dfrac{10}{(0.917...)^{16}} \Rightarrow a = 40$ **2**

It follows that the *exact* value of a is 40. If the value of a is found by calculator, then take care to work with the full available accuracy of the result for r and not the *rounded* value (0.917), given above.

(b) Using $S_n = \dfrac{n}{2}(2a + (n-1)d)$

$10\,000 = \dfrac{n}{2}(2a + 10(n-1)) = n(a + 5(n-1))$

$\Rightarrow a = \dfrac{10\,000}{n} - 5(n-1)$ **2**

n th term is $a + (n-1)d$

$= \dfrac{10\,000}{n} - 5(n-1) + 10(n-1)$

$= \dfrac{10\,000}{n} + 5(n-1)$ **1**

Note: n can only take positive whole number values but the graph of the associated equation

$y = x^2 - 101x + 2000$

$\Rightarrow \dfrac{10\,000}{n} + 5(n-1) < 500$

in continuous variables x and y, can reveal important information about the behaviour of

$n^2 - 101n + 2000.$

$\Rightarrow 10\,000 + 5n^2 - 5n < 500n$ (since $n > 0$)

$\Rightarrow 5n^2 - 505n + 10\,000 < 0$

$\Rightarrow n^2 - 101n + 2000 < 0$

The larger root of the equation
$x^2 - 101x + 2000 = 0$

is given by $x = \dfrac{101 + \sqrt{101^2 - 8000}}{2}$

$= 73.957...$

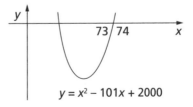

Hence, the largest possible value of n is 73.   **4**

3 (a) When ball strikes floor for third time, total

distance travelled is $1 + 2\left(\frac{3}{5}\right) + 2\left(\frac{3}{5}\right)^2 = \underline{2.92\text{ m}}$ **3**

(b) Total distance travelled cannot exceed

$1 + 2\left(\frac{3}{5}\right) + 2\left(\frac{3}{5}\right)^2 + 2\left(\frac{3}{5}\right)^3 + ... = \dfrac{2}{\left(1 - \frac{3}{5}\right)} - 1$

It is necessary to subtract 1 from the sum of the infinite G.P. because the first term is 1 (*not* 2).

$= \underline{4\text{ m}}.$ **4**

Answer	Mark	Examiner's tip

4 (a) The first four terms of $(1 + kx)^8$ are given by

$$1 + 8kx + \frac{8 \times 7}{2!}(kx)^2 + \frac{8 \times 7 \times 6}{3!}(kx)^3$$

$$= 1 + 8kx + 28k^2x^2 + 56k^3x^3 \qquad \textbf{4}$$

It's a good idea to write the expansion in full first and then simplify it. This is particularly important where minus signs are involved.

Comparing with the given result gives

$8k = 12 \Rightarrow \underline{k = 1.5}, \quad p = 28k^2 \Rightarrow \underline{p = 63} \qquad \textbf{3}$

and $q = 56k^3 \Rightarrow \underline{q = 189}$

The two expansions are equivalent and so corresponding coefficients must be equal.

(b) Coefficient of x^3 is given by $q - p = \underline{126} \qquad \textbf{4}$

5 $(1 + x)^5 = 1 + 5x + 10x^2 + 10x^3 + 5x^4 + x^5 \qquad \textbf{1}$

Letting $x = z + z^2$, the only terms containing z^3 are:

$10x^2 = 10(z + z^2)^2 = 10(z^2 + 2z^3 + z^4)$

and $10x^3 = 10(z + z^2)^3$

$\qquad = 10(z^3 + \dots \text{terms in higher powers})$

The sum of the terms in z^3 is given by

$20z^3 + 10z^3 = 30z^3$.

Hence, the required coefficient of z^3 is 30. $\qquad \textbf{4}$

In this case, the simplest way to obtain the coefficients of the powers of x is to use Pascal's Triangle.

```
            1
          1   1
        1   2   1
      1   3   3   1
    1   4   6   4   1
  1   5  10  10   5   1
```

Note: There is no need to work out the complete expansion of $(1 + z + z^2)^5$. It is only necessary to consider the terms in which z^3 will appear.

6 (i) $(1 - 2x)^{\frac{1}{2}} = 1 + \frac{1}{2}(-2x) + \frac{\frac{1}{2}(-\frac{1}{2})}{2!}(-2x)^2$

$+ \frac{\frac{1}{2}(-\frac{1}{2})(-\frac{3}{2})}{3!}(-2x)^3 + \frac{\frac{1}{2}(-\frac{1}{2})(-\frac{3}{2})(-\frac{5}{2})}{4!}(-2x)^4 + \dots$

$= 1 - x - \frac{x^2}{2} - \frac{x^3}{2} - \frac{5x^4}{8} - \dots$

valid for $|2x| < 1$

i.e. $|x| < \frac{1}{2}$. $\qquad \textbf{6}$

Note: The binomial expansion of $(2 + x)^{\frac{1}{2}}$ for example, could be

worked out as $2^{\frac{1}{2}}\left(1 + \frac{x}{2}\right)^{\frac{1}{2}}$ and

would be valid for $\left|\frac{x}{2}\right| < 1$

i.e for $|x| < 2$.

(ii) $\sqrt{0.8} = (1 - 2 \times 0.1)^{\frac{1}{2}}$

$\approx 1 - 0.1 - \frac{0.1^2}{2} - \frac{0.1^3}{2} - \frac{5 \times 0.1^4}{8}$

$= 0.894\ 437\ 5$

$\Rightarrow \underline{\sqrt{0.8} = 0.8944}$ (correct to four d.p.) $\qquad \textbf{4}$

Note: The series expansion established in part (i) is valid when $x = 0.1$ and so the required result may be obtained by substitution.

This result is easily checked by using the square root function directly.

Answer	Mark	Examiner's tip

7 $(1-4x)^{-\frac{1}{2}} = 1 + (-\frac{1}{2})(-4x) + \frac{(-\frac{1}{2})(-\frac{3}{2})}{2}(-4x)^2$

$\qquad + \frac{(-\frac{1}{2})(-\frac{3}{2})(-\frac{5}{2})}{3!}(-4x)^3 + ...$

$\qquad = 1 + 2x + 6x^2 + 20x^3 + ...$　　**4**

$\dfrac{(1-3x)}{\sqrt{1-4x}} = (1-3x)(1-4x)^{-\frac{1}{2}}$

$\qquad = (1-3x)(1 + 2x + 6x^2 + 20x^3 + ...)$

Considering only the terms in x^3 we have

$20x^3 - 18x^3 = 2x^3$

and so the coefficient of x^3 is 2.　　**2**

Use brackets and include all minus signs before attempting to simplify the result. It is important, here, to appreciate the structure of the question so that the link between the parts is recognised. Questions often follow a pattern in which the results established early on may be **used** later.

Be **clear** about what the question requires you to do. In this case there is no need to work out the full expansion.

8 (a) $t_n = 1 - \dfrac{1}{n}$

as $n \to \infty, \dfrac{1}{n} \to 0$.

$\Rightarrow$ as $n \to \infty, t_n \to 1$

It follows that t_n defines a <u>convergent</u> sequence.　　**1**

In this case, the behaviour of the sequence as n increases is clear, and so the most convincing approach is to deduce the result as shown.

(b) $u_n = 1 - \dfrac{1}{u_{n-1}}$, where $u_1 = 2$

Using the definition,

$u_2 = 1 - \frac{1}{2} = \frac{1}{2}, u_3 = 1 - \dfrac{1}{\left(\frac{1}{2}\right)} = 1 - 2 = -1$

$u_4 = 1 - \dfrac{1}{(-1)} = 1 - (-1) = 2 = u_1$.

It follows that the sequence will continue as:
$2, \frac{1}{2}, -1, 2, \frac{1}{2}, -1, ...$
and so the sequence is <u>periodic</u> (with period 3).　　**3**

It is not immediately apparent just how the sequence will develop and so it's a good idea to apply the definition to find the value of some terms.

The sequence is *divergent*, because the terms do not approach a limit, **but** the most informative description is that it is *periodic* (with period 3) because after every 3 terms the sequence repeats itself.

3 COORDINATE GEOMETRY

Answer	Mark	Examiner's tip

1

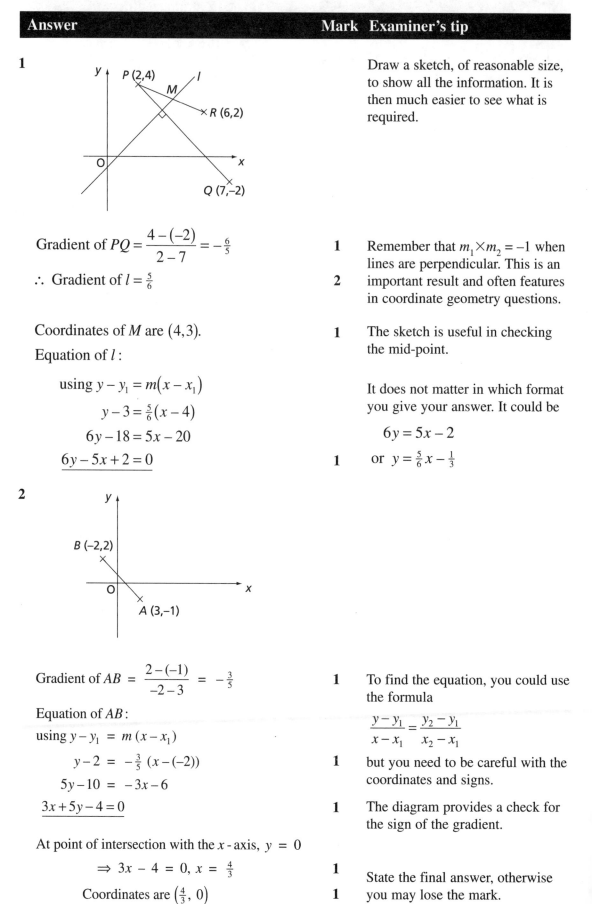

Gradient of $PQ = \dfrac{4-(-2)}{2-7} = -\dfrac{6}{5}$ **1**

∴ Gradient of $l = \dfrac{5}{6}$ **2**

Coordinates of M are $(4,3)$.   **1**

Equation of l:

    using $y - y_1 = m(x - x_1)$

      $y - 3 = \dfrac{5}{6}(x - 4)$

      $6y - 18 = 5x - 20$

      $\underline{6y - 5x + 2 = 0}$ **1**

Tips (right column):
Draw a sketch, of reasonable size, to show all the information. It is then much easier to see what is required.

Remember that $m_1 \times m_2 = -1$ when lines are perpendicular. This is an important result and often features in coordinate geometry questions.

The sketch is useful in checking the mid-point.

It does not matter in which format you give your answer. It could be

      $6y = 5x - 2$

or $y = \dfrac{5}{6}x - \dfrac{1}{3}$

2

Gradient of $AB = \dfrac{2-(-1)}{-2-3} = -\dfrac{3}{5}$ **1**

Equation of AB:

using $y - y_1 = m(x - x_1)$

      $y - 2 = -\dfrac{3}{5}(x - (-2))$ **1**

      $5y - 10 = -3x - 6$

$\underline{3x + 5y - 4 = 0}$ **1**

At point of intersection with the x-axis, $y = 0$

      $\Rightarrow 3x - 4 = 0,\ x = \dfrac{4}{3}$ **1**

    Coordinates are $\left(\dfrac{4}{3},\, 0\right)$ **1**

Tips (right column):
To find the equation, you could use the formula

$$\dfrac{y - y_1}{x - x_1} = \dfrac{y_2 - y_1}{x_2 - x_1}$$

but you need to be careful with the coordinates and signs.

The diagram provides a check for the sign of the gradient.

State the final answer, otherwise you may lose the mark.

Answer	Mark	Examiner's tip

3 $3x + 4y - 36 = 0$

At B, $y = 0 \Rightarrow 3x = 36$, $x = 12$ so B is $(12, 0)$

At C, $x = 0 \Rightarrow 4y = 36$, $y = 9$ so C is $(0, 9)$

(a)

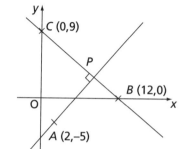

Put all the information obtained so far onto a diagram.

From the diagram:

Gradient of $BC = -\frac{3}{4}$

$\Rightarrow$ Gradient of $AP = \frac{4}{3}$ **1**

The diagram has been useful for obtaining the gradient of BC directly.

Equation of AP:

$$y - y_1 = m(x - x_1)$$
$$y - (-5) = \tfrac{4}{3}(x - 2)$$
$$3y + 15 = 4x - 8$$
$$\Rightarrow \underline{4x - 3y - 23 = 0}$$ **1**

(b) To find the coordinates of P:

P lies on $AP \Rightarrow 4x - 3y - 23 = 0$	(1)	
P lies on $BC \Rightarrow 3x + 4y - 36 = 0$	(2)	
$(1) \times 4$ $16x - 12y - 92 = 0$	(3)	
$(2) \times 3$ $9x + 12y - 108 = 0$	(4)	
$(3) + 4$ $25x \qquad - 200 = 0$		
$x = 8$		

Label each equation for easy reference and explain your working.

Substitute in (2) $24 + 4y - 36 = 0$

$$y = 3$$

$\therefore$ P has coordinates $\underline{(8, 3)}$ **2**

By Pythagoras' Theorem

$$AP^2 = (8 - 2)^2 + (3 - (-5))^2$$
$$= 100$$
$$\Rightarrow AP = 10$$

The perpendicular distance from

A to BC is 10 units   **1**

It is quicker if you recognise the Pythagorean triple 6, 8, 10. Be on the look out for them.

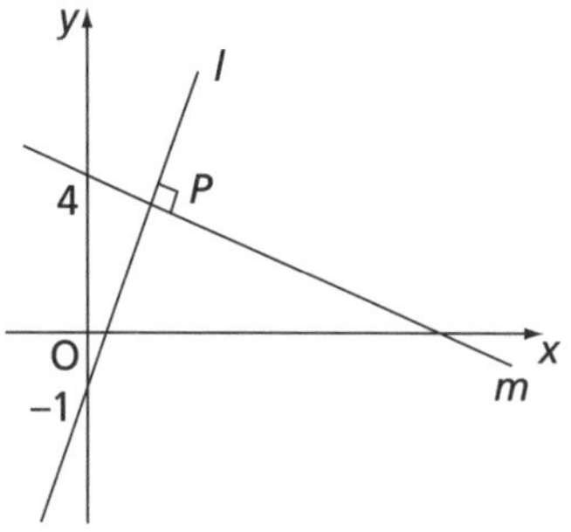

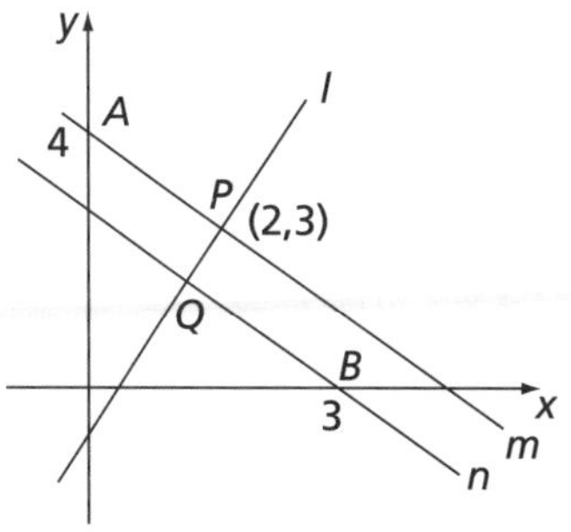

Answer	Mark	Examiner's tip

(c) Area $\triangle ABC = \frac{1}{2} \times CB \times AP$

To find CB you need to focus your attention on a different triangle, $\triangle OCB$.

CB can be found from $\triangle OCB$

$CB^2 = 9^2 + 12^2 = 225 \Rightarrow CB = 15$ **1**

Area $\triangle ABC = \frac{1}{2} \times 15 \times 10$

Another Pythagorean triple has been used!

$= 75$ square units   **1**

4 $l: \ 2x - y - 1 = 0 \Rightarrow y = 2x - 1$

so gradient $= 2$, y-intercept $= -1$ **1**

The format '$y = mx + c$' is useful here, but be careful with notation, since m has been used for the name of the line.

(a) $m:$ gradient $= -\frac{1}{2}$, y-intercept $= 4$

so equation of m is $y = -\frac{1}{2}x + 4$ **2**

At P, $y = 2x - 1$ and $y = -\frac{1}{2}x + 4$

$\therefore 2x - 1 = -\frac{1}{2}x + 4$

$\frac{5}{2}x = 5$

$x = 2$

When $x = 2$, $y = 2(2) - 1 = 3$

$\therefore$ P is $(2,3)$ as required   **1**

The substitution method is a good one to use to solve the simultaneous equations.

(b)

It is helpful to draw another diagram so that you are clear about where the lines and points are.

Equation of n:

$y - y_1 = m(x - x_1)$

$y - 0 = -\frac{1}{2}(x - 3)$

$2y = -x + 3$ **2**

Answer	Mark	Examiner's tip
At Q, $2y = -x + 3$ and $y = 2x - 1$	1	
$\therefore\ 2(2x - 1) = -x + 3$		
$5x = 5$		
$x = 1,\ y = 2(1) - 1 = 1$		
$\therefore\ \underline{Q \text{ has coordinates } (1, 1)}$	1	
(c) $AP^2 = 2^2 + 1^2 = 5 \Rightarrow AP = \sqrt{5}$		Leave the lengths in surd form.
$BQ^2 = 2^2 + 1^2 = 5 \Rightarrow BQ = \sqrt{5}$		
$PQ^2 = 1^2 + 2^2 = 5 \Rightarrow PQ = \sqrt{5}$	3	
$\therefore\ \underline{AP = BQ = PQ}$	1	State the final result, do not leave it to be assumed.
5 (i) $AB = \sqrt{50},\ BC = \sqrt{98},\ AC = \sqrt{148}$		You will need to draw your own diagram showing the coordinates. The lengths can be found using
from which the result follows.	4	
(Check for your values		$d^2 = (x_2 - x_1)^2 + (y_2 - y_1)^2.$
that $AB^2 + BC^2 = CA^2$)		Leave them in surd form.
$A\hat{B}C = 90°$	1	
(ii) Gradient $AB = 1$, gradient $BC = -1$		Use the mid-point formula
Since products of gradients $= -1$,		$\left(\tfrac{1}{2}(x_1 + x_2), \tfrac{1}{2}(y_1 + y_2)\right)$
$A\hat{B}C = 90°$	3	to find M (6, 1) and work out MA and MB.
(iii) $MA = MB = \sqrt{37}$	3	
(iv) $(x - 6)^2 + (y - 1)^2 = 37$	3	Note the instruction to 'write down' which infers that the result is clear from what has already been established.
		Remember that the angle in a semicircle is 90°, so AC is a diameter, M the centre.
6 (a) $x = at,\ y = \dfrac{a}{t}$		
		The curve can be plotted by finding x and y values for various values of t, but it is quicker to spot the cartesian equation. Check on a graphics calculator.
Substituting $t = \dfrac{x}{a}$ into y gives $y = \dfrac{a}{x/a}$		
i.e. $xy = a^2$		
This is the equation of a hyperbola.		
	2	

Answer	Mark	Examiner's tip
(b) $\dfrac{dx}{dt} = a \Rightarrow \dfrac{dt}{dx} = \dfrac{1}{a}$		Remember this standard procedure.
$\dfrac{dy}{dt} = -\dfrac{a}{t^2}$	**2**	
$\dfrac{dy}{dx} = \dfrac{dy}{dt} \times \dfrac{dt}{dx} = \dfrac{1}{a} \times -\dfrac{a}{t^2} = \underline{-\dfrac{1}{t^2}}$	**2**	
(c) At P, $t = 2$ so P is the point $\left(2a, \dfrac{a}{2}\right)$	**1**	Take time to explain what you are doing; do not just write down a series of figures and equations.
At P, $\dfrac{dy}{dx} = -\tfrac{1}{4}$		
$\Rightarrow$ gradient of normal $= 4$	**2**	
Equation of normal at P:		
$y - y_1 = m(x - x_1)$		
$\Rightarrow y - \dfrac{a}{2} = 4(x - 2a)$		
$2y - a = 8x - 16a$		
$\underline{2y = 8x - 15a}$	**2**	
(d) $Q\left(at, \dfrac{a}{t}\right)$ lies on the line		
$2y = 8x - 15a \Rightarrow \dfrac{2a}{t} = 8at - 15a$	**1**	Cancel the factor of a, since $a \neq 0$.
$8t^2 - 15t - 2 = 0$	**1**	
$(8t + 1)(t - 2) = 0$		
$t = -\tfrac{1}{8}, \ t = 2$		Note that $t = 2$ gives the point P, which we know already.
Therefore, at Q, $\underline{t = -\tfrac{1}{8}}$	**2**	

4 FUNCTIONS

Answer	Mark	Examiner's tip

1 $f : x \to \dfrac{1}{2-x} + 3 \quad x \in \mathbb{R}, \ x \neq 2$

(a) $f(5) = \dfrac{1}{2-5} + 3 = -\frac{1}{3} + 3 = 2\frac{2}{3}.$

$ff(5) = f(2\frac{2}{3}) = \dfrac{1}{2 - 2\frac{2}{3}} + 3 = -\frac{3}{2} + 3 = 1\frac{1}{2}.$ **1** $f(3) = 2$ and so $ff(3) = f(2)$ which is not defined since 2 is not in the domain of f.

$ff(k)$ is not defined when $\underline{k = 3}$ **1**

(b) Let $\dfrac{1}{2-x} + 3 = y$

Making x the subject of this equation allows us to identify the reverse process needed for the inverse function.

then $\dfrac{1}{2-x} = y - 3 \ \Rightarrow 2 - x = \dfrac{1}{y-3}$

$\Rightarrow x = 2 - \dfrac{1}{y-3}$ **1** The reverse process is now defined in terms of the variable x.

It follows that $\underline{f^{-1}(x) = 2 - \dfrac{1}{x-3}}.$ **1**

The domain of f^{-1} is given by $x \in \mathbb{R}, \ x \neq 3.$ **1**

2 (a) $k(x) = f(g(x)) = f(3 - 2x) = 2(3 - 2x) - 1$ **1**

$\therefore \underline{k(x) = 5 - 4x}$ **1**

(b) $h(k(x)) = h(5 - 4x) = \frac{1}{4}\big(5 - (5 - 4x)\big)$ **1** Care is needed here with the use of brackets.

$= \frac{1}{4} \times 4x$

$\therefore \underline{h(k(x)) = x}$ **1**

(c) $\underline{h = k^{-1}}$ **1** Note: The wording of part (c) provides a clue to the expected result.

3 (a) $g : x \to x^2 + 1, \quad x \in \mathbb{R}$

The minimum value of $g(x)$ is 1, since $x^2 \geq 0.$

So, the range of g is given by $\underline{\{x : x \geq 1\}}$ **1**

(b) $gf(x) = fg(x) \Rightarrow g(3x - 1) = f(x^2 + 1)$ Care is needed to avoid algebraic errors, particularly in the use of brackets.

$\Rightarrow (3x - 1)^2 + 1 = 3(x^2 + 1) - 1$ **2**

$\Rightarrow 9x^2 - 6x + 2 = 3x^2 + 2$ Don't be tempted to divide throughout by x, or the solution $x = 0$ may be lost.

$\Rightarrow 6x^2 - 6x = 0$ **2**

$\Rightarrow 6x(x - 1) = 0$

$\Rightarrow x = 0 \ \text{or} \ x = 1.$ **1**

Answer	Mark	Examiner's tip

(c) $|f(x)| = 8 \Rightarrow f(x) = 8$ or $f(x) = -8$.　　1

$f(x) = 8 \Rightarrow 3x - 1 = 8 \Rightarrow x = 3$

$f(x) = -8 \Rightarrow 3x - 1 = -8 \Rightarrow x = -2\frac{1}{3}$

hence, the required values are $\underline{3 \text{ and } -2\frac{1}{3}}$　　**2**

Note how the solution branches to take the effect of the modulus function into account.

(d) $h(x) = x^2 + 3x = \left(x + \frac{3}{2}\right)^2 - \frac{9}{4}$

hence, the least value of q is $-\frac{3}{2}$.　　**2**

The required value of q corresponds to the turning point on the graph, since for smaller values h will be many-one.

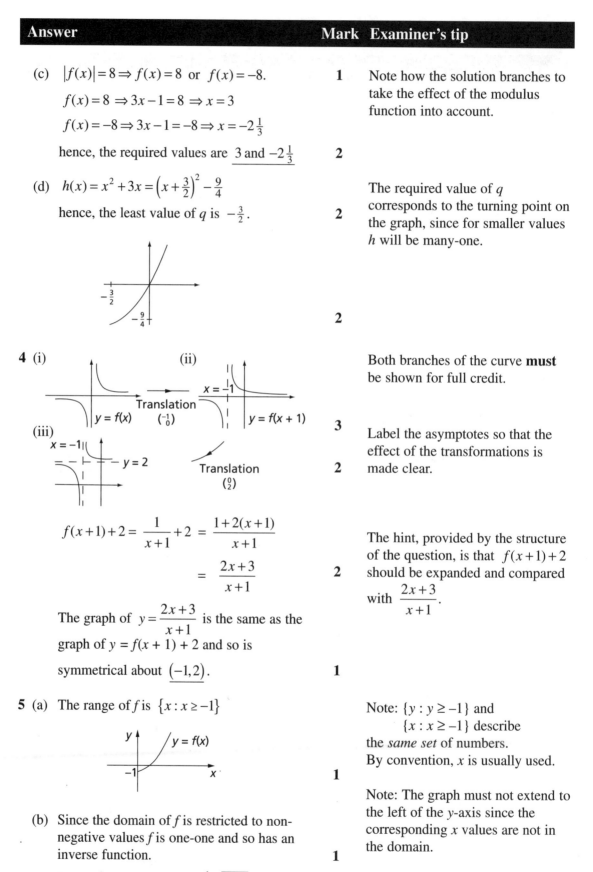

2

4 (i)　　(ii)

3

Both branches of the curve **must** be shown for full credit.

(iii)

2

Label the asymptotes so that the effect of the transformations is made clear.

$f(x+1) + 2 = \dfrac{1}{x+1} + 2 = \dfrac{1 + 2(x+1)}{x+1}$

$= \dfrac{2x+3}{x+1}$　　**2**

The hint, provided by the structure of the question, is that $f(x+1) + 2$ should be expanded and compared with $\dfrac{2x+3}{x+1}$.

The graph of $y = \dfrac{2x+3}{x+1}$ is the same as the graph of $y = f(x+1) + 2$ and so is

symmetrical about $\underline{(-1, 2)}$.　　**1**

5 (a) The range of f is $\{x : x \geq -1\}$

1

Note: $\{y : y \geq -1\}$ and $\{x : x \geq -1\}$ describe the *same set* of numbers. By convention, x is usually used.

(b) Since the domain of f is restricted to non-negative values f is one-one and so has an inverse function.　　**1**

Note: The graph must not extend to the left of the y-axis since the corresponding x values are not in the domain.

Let $4x^2 - 1 = y$ then $x = \frac{1}{2}\sqrt{y+1}$

so $\underline{f^{-1}(x) = \frac{1}{2}\sqrt{x+1}}$　　**2**

Answer	Mark	Examiner's tip

(c) $fg(x) = f\left(\sqrt{(x+6)}\right) = 4(x+6) - 1 = 4x + 23$

$fg(x) \geq f(x) \Rightarrow 4x + 23 \geq 4x^2 - 1$

$\Rightarrow x^2 - x - 6 \leq 0$

$\Rightarrow (x+2)(x-3) \leq 0$ **1**

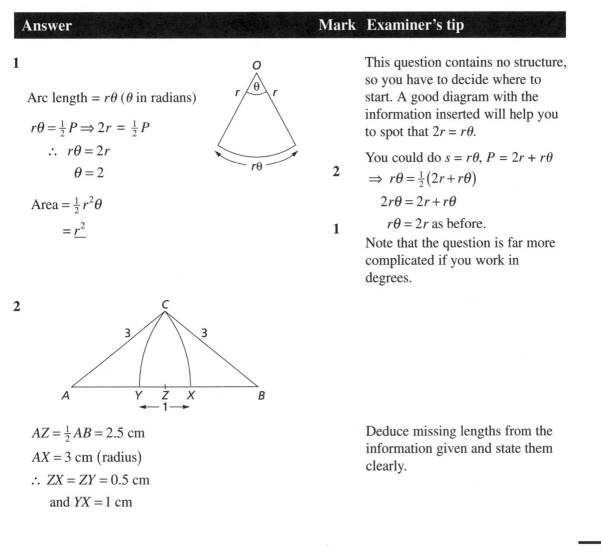

From the sketch, $fg(x) \geq f(x) \Rightarrow 0 \leq x \leq 3$ **1**

An alternative approach is to add the graph of $y = 4x + 23$ to the graph given in part (a). The intersection of the graphs is found by solving $4x + 23 = 4x^2 - 1$ and the solution of $fg(x) \geq f(x)$ is, again, seen to be $0 \leq x \leq 3$.

Care is needed to ensure that only those values of x in the domain are included in the solution.

1 (shown next to the graph)

5 TRIGONOMETRY

Answer	Mark	Examiner's tip

1

Arc length $= r\theta$ (θ in radians)

$r\theta = \frac{1}{2}P \Rightarrow 2r = \frac{1}{2}P$

$\therefore \quad r\theta = 2r$

$\theta = 2$

Area $= \frac{1}{2}r^2\theta$

$= r^2$ **1**

(diagram of sector with centre O, angle θ, radii r, r, and arc $r\theta$)   **2**

This question contains no structure, so you have to decide where to start. A good diagram with the information inserted will help you to spot that $2r = r\theta$.

You could do $s = r\theta$, $P = 2r + r\theta$

$\Rightarrow r\theta = \frac{1}{2}(2r + r\theta)$

$2r\theta = 2r + r\theta$

$r\theta = 2r$ as before.

Note that the question is far more complicated if you work in degrees.

2

(diagram of triangle A, B, C with sides 3, 3 and base points Y, Z, X, with segment 1 marked)

$AZ = \frac{1}{2}AB = 2.5$ cm

$AX = 3$ cm (radius)

$\therefore \quad ZX = ZY = 0.5$ cm

and $YX = 1$ cm

Deduce missing lengths from the information given and state them clearly.

Answer	Mark	Examiner's tip

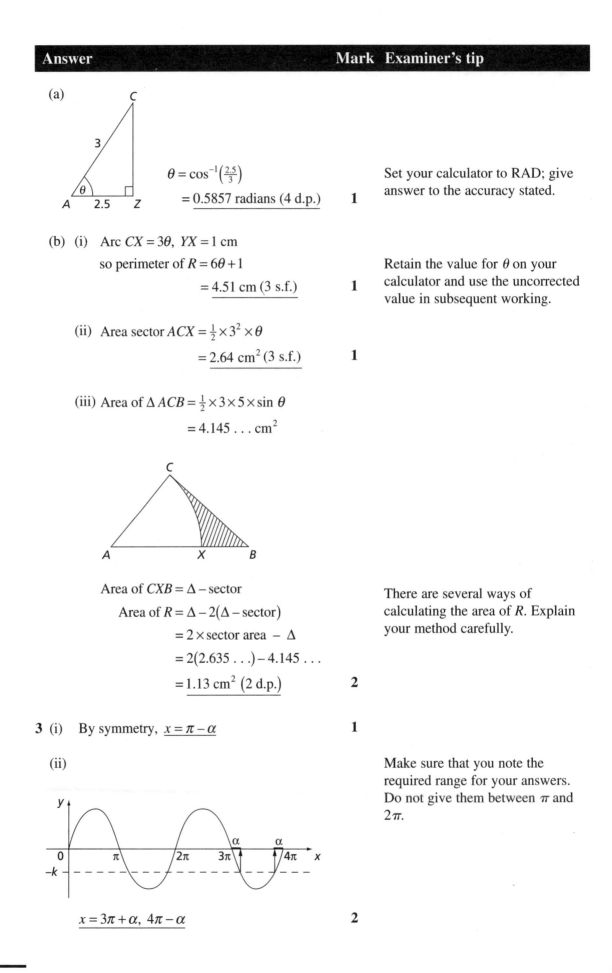

(a)

$$\theta = \cos^{-1}\left(\tfrac{2.5}{3}\right)$$
$$= \underline{0.5857 \text{ radians}} \text{ (4 d.p.)}$$

1

Set your calculator to RAD; give answer to the accuracy stated.

(b) (i) Arc $CX = 3\theta$, $YX = 1$ cm

so perimeter of $R = 6\theta + 1$
$$= \underline{4.51 \text{ cm}} \text{ (3 s.f.)}$$

1

Retain the value for θ on your calculator and use the uncorrected value in subsequent working.

(ii) Area sector $ACX = \tfrac{1}{2} \times 3^2 \times \theta$
$$= \underline{2.64 \text{ cm}^2} \text{ (3 s.f.)}$$

1

(iii) Area of $\triangle ACB = \tfrac{1}{2} \times 3 \times 5 \times \sin\theta$
$$= 4.145\ldots \text{ cm}^2$$

Area of $CXB = \Delta - \text{sector}$

Area of $R = \Delta - 2(\Delta - \text{sector})$
$$= 2 \times \text{sector area} - \Delta$$
$$= 2(2.635\ldots) - 4.145\ldots$$
$$= \underline{1.13 \text{ cm}^2} \text{ (2 d.p.)}$$

2

There are several ways of calculating the area of R. Explain your method carefully.

3 (i) By symmetry, $\underline{x = \pi - \alpha}$

1

(ii)

$$\underline{x = 3\pi + \alpha,\ 4\pi - \alpha}$$

2

Make sure that you note the required range for your answers. Do not give them between π and 2π.

Answer	Mark	Examiner's tip

4 (a)

$$6 \sin^2 x = 5 + \cos x$$

$$6\left(1 - \cos^2 x\right) = 5 + \cos x \qquad \text{1}$$

$$6 - 6 \cos^2 x = 5 + \cos x$$

$$6 \cos^2 x + \cos x - 1 = 0 \qquad \text{1}$$

$$(3 \cos x - 1)(2 \cos x + 1) = 0$$

$$\Rightarrow \ \underline{\cos x = \tfrac{1}{3} \ \text{or} \ \cos x = -\tfrac{1}{2}} \qquad \text{2}$$

This is a common type of question. If you change the format of $\sin^2 x$ you can form a quadratic equation in cos x.

(b) $\cos x = \tfrac{1}{3}$,

principal value $= \cos^{-1}\left(\tfrac{1}{3}\right) = 70.52 \ldots^{\circ}$

$\cos x = -\tfrac{1}{2}$,

principal value $= \cos^{-1}\left(-\tfrac{1}{2}\right) = 120°$

In the range $180° < x < 540°$

The calculator does not give answers in the required range.

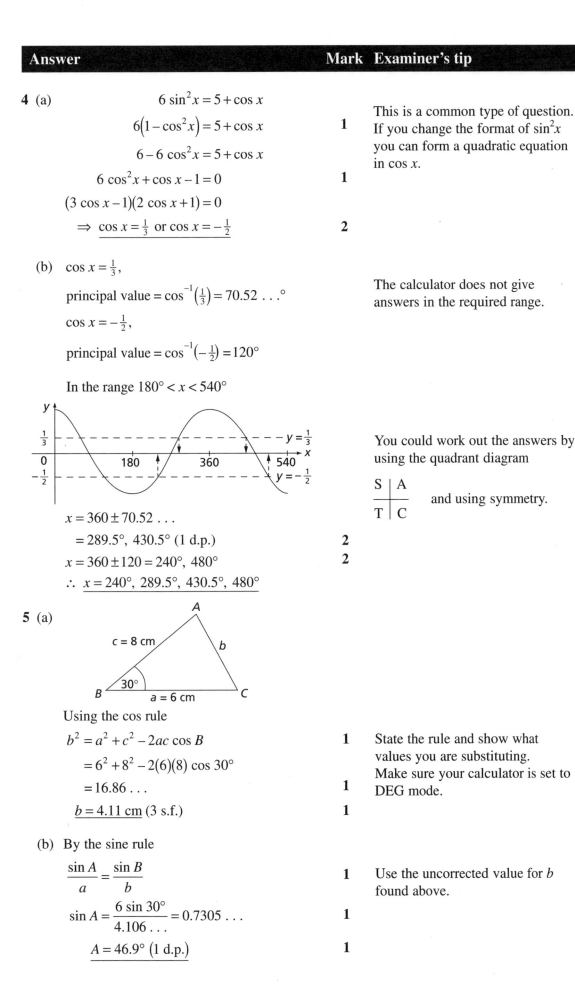

$$x = 360 \pm 70.52 \ldots$$

$$= 289.5°, \ 430.5° \ (1 \ \text{d.p.}) \qquad \text{2}$$

$$x = 360 \pm 120 = 240°, \ 480° \qquad \text{2}$$

$$\therefore \ \underline{x = 240°, \ 289.5°, \ 430.5°, \ 480°}$$

You could work out the answers by using the quadrant diagram

$$\begin{array}{c|c} S & A \\ \hline T & C \end{array} \quad \text{and using symmetry.}$$

5 (a)

Using the cos rule

$$b^2 = a^2 + c^2 - 2ac \cos B \qquad \text{1}$$

$$= 6^2 + 8^2 - 2(6)(8) \cos 30°$$

$$= 16.86 \ldots \qquad \text{1}$$

$$\underline{b = 4.11 \ \text{cm}} \ (3 \ \text{s.f.}) \qquad \text{1}$$

State the rule and show what values you are substituting. Make sure your calculator is set to DEG mode.

(b) By the sine rule

$$\frac{\sin A}{a} = \frac{\sin B}{b} \qquad \text{1}$$

$$\sin A = \frac{6 \sin 30°}{4.106 \ldots} = 0.7305 \ldots \qquad \text{1}$$

$$\underline{A = 46.9°} \ (1 \ \text{d.p.}) \qquad \text{1}$$

Use the uncorrected value for b found above.

Answer	Mark	Examiner's tip

6 (a) $y = 10 - 3\cos kt$

    (i) At low tide, y is least. This occurs when $\cos kt$ is greatest, i.e. when $\cos kt = 1$ so $y_{least} = 10 - 3 = \underline{7\text{ m}}$ **1**

You need to relate the equation to the physical situation.

    (ii) At high tide, y is greatest. This occurs when $\cos kt$ is least, i.e. when $\cos kt = -1$ so $y_{greatest} = 10 + 3 = \underline{13\text{ m}}$ **1**

(b)   **2**

Use the maximum and minimum values established in part (a).

(c) At low tide $\cos kt = 1 \Rightarrow kt = 0 \ \therefore \ t = 0$

    At high tide $\cos kt = -1 \Rightarrow kt = \pi \ \therefore \ t = \dfrac{\pi}{k}$

    $\therefore \ 6.20 = \dfrac{\pi}{k} \Rightarrow k = \dfrac{\pi}{6.20} = \underline{0.51 \ (2\text{ d.p.})}$ **2**

There are other ways of showing this; make your method clear.

7 $\sin 3\theta = \sin(2\theta + \theta)$

    $= \sin 2\theta \cos\theta + \cos 2\theta \sin\theta$

    $= 2\sin\theta \cos\theta \cos\theta + \left(1 - 2\sin^2\theta\right)\sin\theta$

    $= 2\sin\theta \cos^2\theta + \sin\theta - 2\sin^3\theta$

    $= 2\sin\theta\left(1 - \sin^2\theta\right) + \sin\theta - 2\sin^3\theta$

    $= 2\sin\theta - 2\sin^3\theta + \sin\theta - 2\sin^3\theta$

    $= \underline{3\sin\theta - 4\sin^3\theta}$ **3**

Knowledge of the use of trig identities is needed here. Use your formulas booklet to check your accuracy, so that you can progress confidently. Notice that the result contains only terms in $\sin\theta$.

    $\sin 3\theta = 2\sin\theta$

Refer to first part of question.

    $\Rightarrow 3\sin\theta - 4\sin^3\theta = 2\sin\theta$

    $\sin\theta - 4\sin^3\theta = 0$

    $\sin\theta\left(1 - 4\sin^2\theta\right) = 0$

    $\sin\theta\,(1 - 2\sin\theta)(1 + 2\sin\theta) = 0$ **2**

    $\therefore \ \sin\theta = 0 \ \text{or} \ \sin\theta = \pm\tfrac{1}{2}$ **1**

There are several traps that you might fall into. If you 'cancel' $\sin\theta$, you will lose some solutions. Also you should factorise the equation fully so that you realise that $\sin\theta = \tfrac{1}{2}$ or $\sin\theta = -\tfrac{1}{2}$.

    $\sin\theta = 0 \Rightarrow \theta = 0, \ 180°, \ 360°$

    $\sin\theta = \tfrac{1}{2} \Rightarrow \theta = 30°, \ 150°$ **1**

    $\sin\theta = -\tfrac{1}{2} \Rightarrow \theta = 210°, \ 330°$ **1**

If you solve $1 - 4\sin^2\theta = 0$ by writing it as $\sin^2\theta = \tfrac{1}{4}$ you may forget that there are two square roots.

    $\underline{\theta = 0, \ 30°, \ 150°, \ 180°, \ 210°, \ 330°, \ 360°}$

Answer	Mark	Examiner's tip

8 (a) $f(x) = \cos x° - \tan x°$

Now $\cos x°$ is defined for all values of x in the range -180 to 360, but $\tan x°$ is not defined when $x = -90, 90$ or 270

∴ $f(x)$ is not defined for $x = \pm 90, 270$ **1**

Knowledge of the basic trigonometric curves is required.

(b)

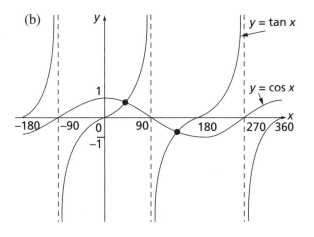

If you are using your graphics calculator, choose a suitable scale.

Show the scale on the x-axis, labelling the points where the curve cross it.   **1**

(c) Now $f(x) = 0$ when $\cos x° - \tan x° = 0$

i.e. $\cos x° = \tan x°$

The values of x which satisfy this equation are given by the x coordinates of the points of intersection of the curves. From the sketch, we see that there are 2 roots.   **1**

You must justify your answer for the number of roots.

(d) There is one root between 0 and 90, and another between 90 and 180.

Consider $f(x) = \cos x° - \tan x°$,

we want $f(x) = 0$

You need to use trial and improvement methods, homing in on the required values.

If you are using the trace facility on the calculator, you must show some working to justify your answers.

$f(30) = 0.288 \ldots$ $f(120) = 1.23 \ldots$
$f(40) = -0.073 \ldots$ $f(130) = 0.54 \ldots$
$f(35) = 0.118 \ldots$ $f(140) = 0.07 \ldots$
$f(38) = 0.006 \ldots$ $f(150) = -0.288 \ldots$
$f(39) = -0.03 \ldots$ $f(141) = 0.03 \ldots$
                              $f(142) = -0.006 \ldots$

Look for a sign change.
See Numerical methods, Unit 8.

Therefore   Therefore
$38 < x < 39$ $141 < x < 142$ **2**

Answer	Mark	Examiner's tip
9 $4\tan^2 x + 12\sec x + 1 = 0$		Remember the relationship between sec x and tan x: $1 + \tan^2 x = \sec^2 x$.
$4(\sec^2 x - 1) + 12\sec x + 1 = 0$	1	
$4\sec^2 x + 12\sec x - 3 = 0$	1	
$\sec x = \dfrac{-12 \pm \sqrt{144 - 4(4)(-3)}}{8}$		The quadratic formula is required here, since the equation does not factorise.
$= \dfrac{-12 \pm \sqrt{192}}{8}$		
$\therefore \ \sec x = -3.232\ldots \Rightarrow \cos x = -0.3094\ldots$		
$x = \pm 108°$ (nearest degree)	2	Give your answers in the range specified.
or $\sec x = 0.232\ldots \Rightarrow \cos x = 4.309\ldots$		
no possible solutions		
$\therefore \ \underline{x = \pm 108°}$	2	
10 (a) $y = a + b \sin cx$		'Write down' infers that you should be able to spot the values straight away. You do not need to put supporting reasons.
$\underline{a = 2,\ b = 1,\ c = 2}$	3	
(b) $\quad 2 + \sin 2x = 2.5$		In fact $y = \sin x$ has been stretched by factor $\frac{1}{c}$ from $x = 0$, stretched by factor b from $y = 0$ and translated by $\begin{pmatrix} 0 \\ a \end{pmatrix}$ – see Unit 4, Functions.
$\sin 2x = 0.5$		
In given range, $2x = \dfrac{\pi}{6},\ \dfrac{5\pi}{6}$		
$\Rightarrow \underline{x = \dfrac{\pi}{12},\ \dfrac{5\pi}{12}}$	3	Give your answers to (b) in terms of π.

6 DIFFERENTIATION

Answer	Mark	Examiner's tip
1 $y = \dfrac{1}{x^3} + \cos 3x = x^{-3} + \cos 3x$		Change the format of y so that it is easier to differentiate.
$\dfrac{dy}{dx} = -3x^{-4} - 3\sin 3x$		
$= \underline{-\dfrac{3}{x^4} - 3\sin 3x}$	4	

Answer	Mark	Examiner's tip

2 $r = \dfrac{1+4t}{2+t}$

(a) When $t = 0$, $r = \frac{1}{2}$ **1**   The expression 'initial' relates to $t = 0$. You need to realise this, otherwise it is difficult to progress.

r is double its original value when $r = 1$

$$\Rightarrow 1 = \frac{1+4t}{2+t}$$ **1**

$$2 + t = 1 + 4t$$

$$3t = 1$$

$$t = \tfrac{1}{3}$$ **1**

<u>The radius doubles its initial value</u>

<u>in $\frac{1}{3}$ sec.</u>

(b) $\dfrac{dr}{dt} = \dfrac{(2+t)\,4 \;-\; (1+4t)\,1}{(2+t)^2}$ **2**   The rate of increase of the radius, in cm s^{-1}, is $\dfrac{dr}{dt}$. You need to use the quotient rule to find it.

$$= \frac{7}{(2+t)^2}$$

When $t = 3$, $\dfrac{dr}{dt} = \dfrac{7}{25} = 0.28$ **1**

<u>The rate of increase of the radius</u>   **2**   State your answer clearly, remembering the units, or you may lose marks.

<u>when $t = 3$ is 0.28 cm s^{-1}.</u>

(c) $r = \dfrac{1+4t}{2+t} = \dfrac{\dfrac{1}{t}+4}{\dfrac{2}{t}+1}$ You need to change the form of the expression for r. You could divide to give $r = 4 - \dfrac{7}{t+2}$.

As $t \to \infty$, $\dfrac{1}{t} \to 0$ $\therefore$ $\underline{r \to 4}$.   **2**

3 $ky = a^x$

(7, 12) lies on the curve, so $12k = a^7$ (1)   The simplest way to eliminate k from these equations is to divide.

(12, 7) lies on the curve, so $7k = a^{12}$ (2)   **2**

$(2) \div (1)$ gives $a^5 = \frac{7}{12} \Rightarrow \underline{a = 0.90\ (2\ \text{s.f.})}$ **2**   Note the instruction to give the value to 2 s.f.

Substituting in (1) gives

                          Use all the figures on your calculator for a, *not* the 2 s.f. value.

$$k = \frac{a^7}{12} = \underline{0.039\ (2\ \text{s.f.})}$$ **2**

Answer	Mark	Examiner's tip

$ky = a^x \Rightarrow \ln(ky) = \ln(a^x)$

$\ln k + \ln y = x \ln a$

$\ln y = x \ln a - \ln k$

$\dfrac{1}{y} \dfrac{dy}{dx} = \ln a$

$\dfrac{dy}{dx} = y \ln a$

$= \dfrac{a^x \ln a}{k}$ **3**

When $x = 20$, $\dfrac{dy}{dx} = \dfrac{a^{20} \ln a}{k} = \underline{-0.3 \ (1 \text{ d.p.})}$ **2**

The derivative of a^x may be given to you as a standard result in your formulae booklet, in which case it can be quoted:

$\dfrac{d}{dx}(a^x) = a^x \ln a$

The method shown here involves implicit differentiation.

Give the value to the required accuracy.

---

4 $y = x + \dfrac{4}{x} = x + 4x^{-1}$

$\dfrac{dy}{dx} = 1 - 4x^{-2}$

$= 1 - \dfrac{4}{x^2}$ **1**

At stationary points, $\dfrac{dy}{dx} = 0$

so $1 - \dfrac{4}{x^2} = 0 \Rightarrow x^2 = 4$

$x = \pm 2$

When $x = 2$, $y = 2 + \frac{4}{2} = 4$ **1**

When $x = -2$, $y = -2 - \frac{4}{2} = -4$ **1**

<u>There are stationary points at</u>

<u>$(2, 4)$ and $(-2, -4)$.</u>

Now $\dfrac{d^2y}{dx^2} = \dfrac{8}{x^3}$

When $x = 2$, $\dfrac{d^2y}{dx^2} > 0 \Rightarrow$ minimum point   **1**

When $x = -2$, $\dfrac{d^2y}{dx^2} < 0 \Rightarrow$ maximum point   **1**

<u>$(2, 4)$ is a maximum point and</u>

<u>$(-2, -4)$ is a minimum point.</u>

The first step is to rewrite the term $\dfrac{4}{x}$ as $4x^{-1}$ in order to apply the rule for differentiation. Often the need arises to change the form of some given information.

Note that the question asks for stationary points; find more than one.

You could consider y values near the point or consider the sign of $\dfrac{dy}{dx}$ near the point – but do not choose $x = 0$ since the function is undefined. Make your method clear and state your conclusions.

Try plotting the curve on a graphic calculator to confirm your conclusions.

Answer	Mark	Examiner's tip

If y increases as x increases, then $\dfrac{dy}{dx} > 0$

1   Take care with inequalities. It is possible to multiply by x^2 here, since $x^2 > 0$.

i.e. $1 - \dfrac{4}{x^2} > 0 \Rightarrow 1 > \dfrac{4}{x^2} \Rightarrow x^2 > 4$

Do not forget negative values. In more complicated situations, sketch the function.

$\therefore$ $\underline{x < -2 \text{ or } x > 2}$ **2**

5 $y = 4x - x^2 \Rightarrow \dfrac{dy}{dx} = 4 - 2x$ **1**

When $x = 0$, $\dfrac{dy}{dx} = 4$ **1**

1   Remember that the gradient is given by the tan of the angle between the line and the positive x-axis.

$\tan(A - B) = \dfrac{\tan A - \tan B}{1 + \tan A \tan B}$

The required angle is $A - B$.

$\qquad = \dfrac{4 - 1}{1 + 4(1)} = 0.6$ **2**

$\therefore A - B = 31.0° \text{ (1 d.p.)}$ **1**

Therefore the angle between the line
$\underline{y = x \text{ and the tangent is } 31.0° \text{ (1 d.p.)}}$

6 $y = 0$ when $80t^2 = 0$

$\qquad\qquad 8t(10 - t) = 0$

$\qquad\qquad\quad t = 0 \text{ or } t = 10$

The question indicates that there is more than one point, so do not be hasty and just say that $t = 10$.

$\dfrac{dx}{dt} = 160 - 12t \Rightarrow \dfrac{dt}{dx} = \dfrac{1}{160 - 12t}$

$\dfrac{dy}{dt} = 80 - 16t$ **1**   State your working clearly.

Answer	Mark	Examiner's tip

Now $\dfrac{dy}{dx} = \dfrac{dy}{dt} \times \dfrac{dt}{dx}$ **1**

$\qquad = \dfrac{80 - 16t}{160 - 12t}$ **1**   There is no need to simplify here.

When $t = 0$, $\dfrac{dy}{dx} = \dfrac{80}{160} = \underline{0.5}$

When $t = 10$, $\dfrac{dy}{dx} = \dfrac{80 - 160}{160 - 120} = \underline{-2}$ **2**

7 (a) $X = 500\, e^{-\frac{1}{5}t}$

$200 = 500\, e^{-\frac{1}{5}t} \Rightarrow e^{-\frac{1}{5}t} = \frac{200}{500} = 0.4$ (1)   **2**

$\qquad\qquad\qquad -\frac{1}{5}t = \ln 0.4$ Take logs to base e of both sides.

$\qquad\qquad\qquad\quad t = -5 \ln 0.4$

$\qquad\qquad\qquad\qquad = \underline{4.58\ \text{hr (2 d.p.)}}$ **1**

(b) (i) $\dfrac{dX}{dt} = -\frac{1}{5}\left(500\, e^{-\frac{1}{5}t}\right)$

$\qquad\quad = \underline{-100\, e^{-\frac{1}{5}t}}$ **2**   Notice that you have a value for $e^{-\frac{1}{5}t}$ from part (a). There is no need to substitute the value of t found in (a), but if you do, use the uncorrected value, *not* the 2 d.p. answer.

(ii) When $X = 200$, $e^{-\frac{1}{5}t} = 0.4$ from (1)

$\qquad \dfrac{dX}{dt} = -100 \times 0.4$

$\qquad\qquad = -40$

This indicates that X is decreasing, and <u>the rate of decrease is 40 milligrammes per hour.</u>   **1**

8 $y = x^3 + bx^2 + cx$

(a) $\dfrac{dy}{dx} = 3x^2 + 2bx + c$ **2**

(b)   When $x = -1$ and $x = 3$, $\dfrac{dy}{dx} = 0$

$x = -1 \Rightarrow\ 0 = 3 - 2b + c$ (1)   **2**   Explain how you are solving the simultaneous equations.

$x = 3 \ \Rightarrow\ 0 = 27 + 6b + c$ (2)

$(2) - (1) \quad 0 = 24 + 8b$

$\qquad\qquad\quad b = -3$ **1**

Substitute in (1) $\underline{c = -9}$ **2**

Answer	Mark	Examiner's tip

(c) $y = x^3 - 3x^2 - 9x$

When $x = -1$, $y = (-1)^3 - 3(-1)^2 - 9(-1) = 5$

When $x = 3$, $y = 3^3 - 3(3^2) - 9(3) = -27$ **2**

$\therefore$ The local maximum value of y is 5,

the local minimum value of y is -27.

(d) A zero is where the function $y = f(x)$ intersects $y = 0$. The translated graph will have only one zero when $d > 27$ or $d < -5$.   **2**

You may find it helpful to draw sketches to explain your answers.

9 (a) $y = 2x^{-\frac{1}{3}} + x^{\frac{2}{3}}$

$\dfrac{dy}{dx} = \left(-\frac{1}{3}\right) 2x^{-\frac{4}{3}} + \frac{2}{3}x^{-\frac{1}{3}}$ **3**

At $x = 1$, $\dfrac{dy}{dx} = -\frac{2}{3} + \frac{2}{3} = 0$ as required   **2**

Take time to work out the indices properly, remembering that you need to reduce the power of x by 1 when differentiating.

(b) $A = \displaystyle\int y\, dx$

so $A = \displaystyle\int_1^8 \left(2x^{-\frac{1}{3}} + x^{\frac{2}{3}}\right) dx$ **2**

$= \left[\dfrac{2}{\frac{2}{3}} x^{\frac{2}{3}} + \dfrac{x^{\frac{5}{3}}}{\frac{5}{3}} \right]_1^8$

$= \left[3x^{\frac{2}{3}} + \frac{3}{5} x^{\frac{5}{3}} \right]_1^8$ **3**

$= 3(8)^{\frac{2}{3}} + \frac{3}{5}(8)^{\frac{5}{3}} - \left(3 + \frac{3}{5}\right)$

$= 27.6$

The area of R is 27.6 square units.   **2**

A common feature of calculus questions is to include differentiation and integration in the same question.
Be clear about each process.
When integrating, raise the power of x by 1.

For more practice see the integration questions.

7 INTEGRATION

Answer	Mark	Examiner's tip

1 (a) $\dfrac{\mathrm{d}}{\mathrm{d}x}\left(1+x^3\right)^{\frac{1}{2}} = \tfrac{1}{2}\left(1+x^3\right)^{-\frac{1}{2}} \times 3x^2$

$$= \dfrac{3x^2}{2\sqrt{1+x^3}}$$

Mark: **3**

Examiner's tip: Use the chain rule carefully.

(b) $\displaystyle\int_0^2 \dfrac{x^2}{\sqrt{1+x^3}}\,\mathrm{d}x = \tfrac{2}{3}\left[\sqrt{1+x^3}\right]_0^2$

Mark: **2**

$$= \tfrac{2}{3}\left(\sqrt{9}-\sqrt{1}\right)$$

$$= \underline{\tfrac{4}{3}}$$

Mark: **2**

Examiner's tip: The wording of the question directs you to look for a relationship between the integral and the function just differentiated. A suitable substitution would be $1+x^3 = u^2$.

2

$$C_1 : y = \dfrac{1}{x}$$

$$C_2 : y = kx^2$$

Examiner's tip: It is a good idea to draw a sketch. Both curves are standard ones which you should know.

(a) At P, $\dfrac{1}{x} = kx^2$

Mark: **1**

$$\Rightarrow k = \dfrac{1}{x^3}$$

Since $x = \tfrac{1}{2}$, $k = \dfrac{1}{\left(\tfrac{1}{2}\right)^3} = 8$

Mark: **1**

Examiner's tip: You are effectively solving two simultaneous equations.

(b) $C_1 : y = x^{-1}$

$$\dfrac{\mathrm{d}y}{\mathrm{d}x} = -\dfrac{1}{x^2}$$

Mark: **2**

At P, $x = \tfrac{1}{2} \Rightarrow \dfrac{\mathrm{d}y}{\mathrm{d}x} = -\dfrac{1}{\left(\tfrac{1}{2}\right)^2} = -4$

Mark: **1**

Examiner's tip: You need the gradient at a particular point, so substitute the x-value at that point.

(c)

Area $A = \displaystyle\int_0^{\frac{1}{2}} 8x^2\,\mathrm{d}x$

$$= \tfrac{8}{3}\left[x^3\right]_0^{\frac{1}{2}}$$

Mark: **2**

$$= \underline{\tfrac{1}{3}}$$

Mark: **2**

Examiner's tip: Find the required area in two parts. The sketch will help you to clarify which curve you need for each part and the limits required.

Answer	Mark	Examiner's tip

Area $B = \int_{\frac{1}{2}}^{2} \frac{1}{x} \, dx$ — 1

$= \left[\ln |x| \right]_{\frac{1}{2}}^{2}$ — 1

$= \ln 2 - \ln \frac{1}{2}$

There is no need to find the area of B numerically until the final line.

$= \ln 2 - \ln(2^{-1})$

$= 2 \ln 2$ — 2

Total area $= \frac{1}{3} + 2 \ln 2$

Give the answer to the requested number of decimal places.

$= 1.72 \ (2 \ \text{d.p.})$ — 2

3 (a)

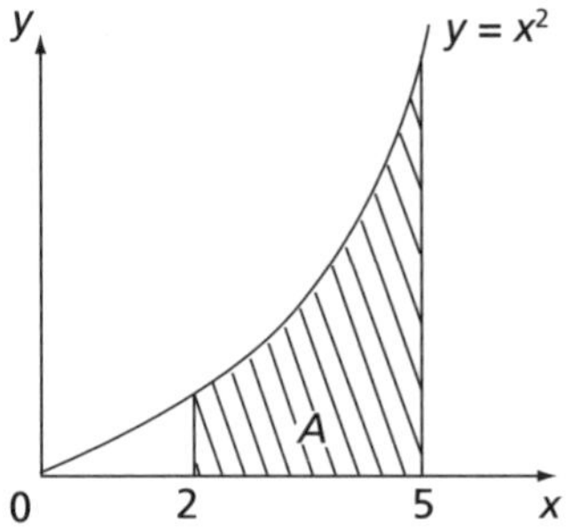

Note that you are instructed to draw a sketch; do not forget to do so. — 1

Area $A = \int y \, dx$

$= \int_{2}^{5} x^2 \, dx$

$= \left[\frac{x^3}{3} \right]_{2}^{5}$

$= 39 \ \text{square units}$ — 1

(b)

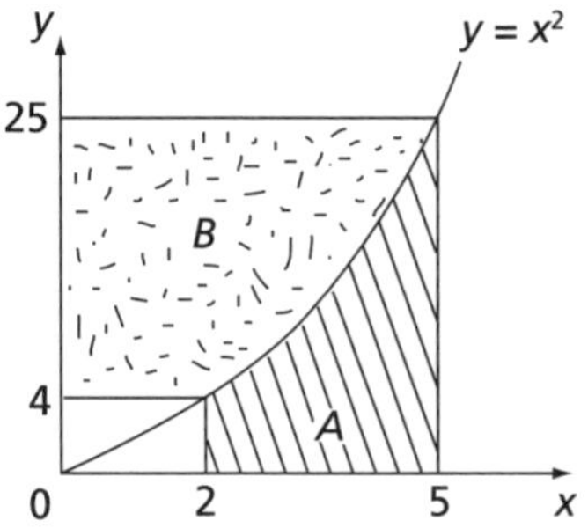

A clear diagram is essential here to explain your method.

Now $B = \int x \, dy$

$= \int_{4}^{25} \sqrt{y} \, dy$ — 1

To find area B, consider strips parallel to the x-axis and use $\int x \, dy$.

Answer	Mark	Examiner's tip

From the diagram

Area $B = 5 \times 25 - \text{Area } A - 4 \times 2$

$\qquad = 125 - 39 - 8$

$\qquad = 78$ — 1 — Be ready to think about the physical interpretation of the integral, which in this case is the area enclosed by the curve, the y-axis, $y = 4$ and $y = 25$. Note the firm instruction about relating part (b) to part (a).

Therefore $\int_4^{25} \sqrt{y}\; \mathrm{d}y = 78$

(c)

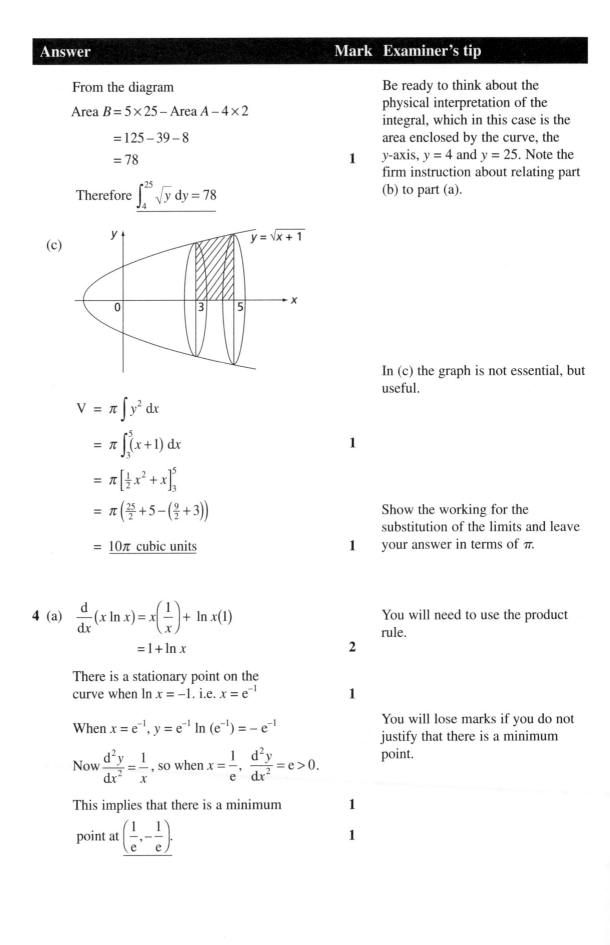

In (c) the graph is not essential, but useful.

$V = \pi \int y^2\, \mathrm{d}x$

$\quad = \pi \int_3^5 (x+1)\, \mathrm{d}x$ — 1

$\quad = \pi \left[\tfrac{1}{2}x^2 + x \right]_3^5$

$\quad = \pi \left(\tfrac{25}{2} + 5 - \left(\tfrac{9}{2} + 3 \right) \right)$ — Show the working for the substitution of the limits and leave your answer in terms of π.

$\quad = \underline{10\pi \text{ cubic units}}$ — 1

4 (a) $\dfrac{\mathrm{d}}{\mathrm{d}x}(x \ln x) = x \left(\dfrac{1}{x} \right) + \ln x (1)$ — You will need to use the product rule.

$\qquad\qquad = 1 + \ln x$ — 2

There is a stationary point on the curve when $\ln x = -1$. i.e. $x = e^{-1}$ — 1

When $x = e^{-1}$, $y = e^{-1} \ln (e^{-1}) = -e^{-1}$ — You will lose marks if you do not justify that there is a minimum point.

Now $\dfrac{\mathrm{d}^2 y}{\mathrm{d}x^2} = \dfrac{1}{x}$, so when $x = \dfrac{1}{e}$, $\dfrac{\mathrm{d}^2 y}{\mathrm{d}x^2} = e > 0$.

This implies that there is a minimum — 1

point at $\left(\dfrac{1}{e}, -\dfrac{1}{e} \right)$. — 1

Answer	Mark	Examiner's tip

(b) Since $\dfrac{d^2y}{dx^2} = \dfrac{1}{x}$ and $x > 0$, $\dfrac{d^2y}{dx^2}$ is never
zero so the curve has no point of inflection.   **1**

At a point of inflection,
$$\dfrac{d^2y}{dx^2} = 0$$

(c)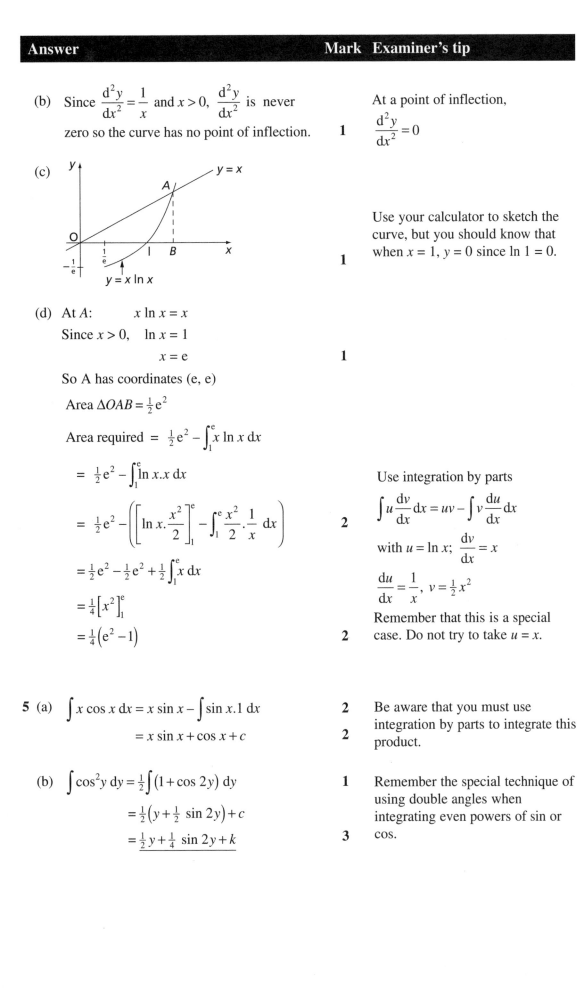

1

Use your calculator to sketch the curve, but you should know that when $x = 1$, $y = 0$ since $\ln 1 = 0$.

(d) At A: $x \ln x = x$

Since $x > 0$, $\ln x = 1$
$$x = e \qquad \textbf{1}$$

So A has coordinates (e, e)

Area $\triangle OAB = \frac{1}{2}e^2$

Area required $= \frac{1}{2}e^2 - \int_1^e x \ln x \, dx$

$\quad = \frac{1}{2}e^2 - \int_1^e \ln x . x \, dx$

$\quad = \frac{1}{2}e^2 - \left(\left[\ln x . \dfrac{x^2}{2} \right]_1^e - \int_1^e \dfrac{x^2}{2} . \dfrac{1}{x} \, dx \right)$ **2**

$\quad = \frac{1}{2}e^2 - \frac{1}{2}e^2 + \frac{1}{2} \int_1^e x \, dx$

$\quad = \frac{1}{4} \left[x^2 \right]_1^e$

$\quad = \frac{1}{4} \left(e^2 - 1 \right)$ **2**

Use integration by parts
$$\int u \dfrac{dv}{dx} dx = uv - \int v \dfrac{du}{dx} dx$$
with $u = \ln x$; $\dfrac{dv}{dx} = x$

$\dfrac{du}{dx} = \dfrac{1}{x}$, $v = \frac{1}{2}x^2$

Remember that this is a special case. Do not try to take $u = x$.

5 (a) $\displaystyle\int x \cos x \, dx = x \sin x - \int \sin x .1 \, dx$ **2**

$\qquad\qquad = x \sin x + \cos x + c$ **2**

Be aware that you must use integration by parts to integrate this product.

(b) $\displaystyle\int \cos^2 y \, dy = \frac{1}{2} \int (1 + \cos 2y) \, dy$ **1**

$\qquad = \frac{1}{2} \left(y + \frac{1}{2} \sin 2y \right) + c$

$\qquad = \frac{1}{2} y + \frac{1}{4} \sin 2y + k$ **3**

Remember the special technique of using double angles when integrating even powers of sin or cos.

Letts
Q&A

Answer	Mark	Examiner's tip

$$\int \frac{1}{\sec^2 2y}\, dy = \int x \cos x\, dx$$

$$\Rightarrow \int \cos^2 2y\, dy = \int x \cos x\, dx \qquad 1$$

$$\tfrac{1}{2}\int (1 + \cos 4y)\, dy = x \sin x + \cos x + c$$

$$\tfrac{1}{2}\left(y + \tfrac{1}{4}\sin 4y\right) = x \sin x + \cos x + c$$

$$\therefore\ y + \tfrac{1}{4}\sin 4y = 2(x \sin x + \cos x) + d \qquad 3$$

Separate the variables carefully. Parts (a) and (b) have already prepared you for the integration of each side, but note that you must alter the format of $\frac{1}{\sec^2 2y}$ and then use a similar technique to part (b) to integrate it.

6 (a) We are given $\dfrac{dR}{dt} = -kR,\ k>0$

$$\int \frac{1}{R}\, dR = -k \int dt$$

$$\ln R = -kt + c$$

$$\Rightarrow R = e^{-kt+c} = e^{-kt}.e^c = Ae^{-kt}$$

Now when $t = 0,\ R = 10$, so $10 = A$

$$\therefore\ \underline{R = 10e^{-kt}} \qquad 2$$

This is a very common differential equation and it is useful to be familiar with this form of the solution, in which the integration constant A is equal to the initial value of the variable, i.e. the value when $t = 0$.

(b) When $t = 1600,\ R = 5$

$$5 = 10e^{-1600\,k} \qquad 1$$

$$\tfrac{1}{2} = e^{-1600\,k}$$

$$2 = e^{1600\,k}$$

$$\ln 2 = 1600\,k$$

$$\Rightarrow\ k = \frac{\ln 2}{1600} \qquad 1$$

$\tfrac{1}{2} = 2^{-1}$, so $2^{-1} = e^{-1600k}$

$\Rightarrow 2 = e^{1600k}$

Your method of working must be shown.

(c) When $t = 100,\ R = 10e^{-\frac{\ln 2}{1600}(100)} \qquad 1$

$$= 9.58 \text{ grams } (2 \text{ d.p.}) \qquad 1$$

Again, show this line. If you just write down the numerical answer you will lose 2 marks.

7 (i) $\dfrac{dr}{dt} = k \Rightarrow r = kt + c \quad (k>0)$

$t = 0,\ r = 1 \Rightarrow 1 = c,\ \therefore r = kt + 1$

$t = 10,\ r = 2 \Rightarrow 2 = 10k + 1,\ \therefore k = 0.1$

The model is $r = 0.1t + 1$, and the rate of increase is 0.1

$t = T,\ r = 4 \Rightarrow 4 = 0.1T + 1$

$$\underline{T = 30} \qquad 3$$

Do not forget to include the integration constant, then use the two conditions to find k and c; k gives the rate of increase.

Letts Q&A

Answer	Mark	Examiner's tip

(ii) $\dfrac{\mathrm{d}r}{\mathrm{d}t} = \dfrac{k}{r}$ **1**

$\displaystyle\int r\,\mathrm{d}r = \int k\,\mathrm{d}t$ **1**   Separate the variables and integrate both sides.

$\frac{1}{2}r^2 = kt + c$ **2**

$t = 0,\ r = 1 \Rightarrow \frac{1}{2} = c \quad \therefore \frac{1}{2}r^2 = kt + \frac{1}{2}$ **1**

$\qquad\qquad\qquad\qquad r^2 = 2kt + 1$

$t = 10,\ r = 2 \Rightarrow 4 = 20k + 1$

$\qquad\qquad\qquad k = 0.15$ **2**

The model is $r^2 = 0.3t + 1$ It helps to state the formula clearly, so that you can use it in the last part.

$t = T,\ r = 4 \Rightarrow\ 16 = 0.3T + 1$

$\qquad\qquad\qquad \underline{T = 50}$ **1**

8 Let $x = 2\cos\theta \Rightarrow \dfrac{\mathrm{d}x}{\mathrm{d}\theta} = -2\sin\theta$ **1**

and $\sqrt{4 - x^2} = \sqrt{4 - 4\cos^2\theta} = 2\sin\theta$ Change the limits to those of the new variable.

Limits: when $x = 1,\ \cos\theta = \frac{1}{2} \Rightarrow \theta = \dfrac{\pi}{3}$ You should recognise the values of θ here and leave them as multiples of π.

$\qquad\quad$ when $x = \sqrt{2},\ \cos\theta = \dfrac{\sqrt{2}}{2} \Rightarrow \theta = \dfrac{\pi}{4}$ **1**

$I = \displaystyle\int_{\frac{\pi}{3}}^{\frac{\pi}{4}} -\dfrac{1}{4\cos^2\theta . 2\sin\theta} \cdot 2\sin\theta\ \mathrm{d}\theta$ Writing $\dfrac{1}{\cos^2\theta}$ as $\sec^2\theta$ leads to immediate recognition of the integral.

$= -\frac{1}{4}\displaystyle\int_{\frac{\pi}{3}}^{\frac{\pi}{4}} \sec^2\theta\ \mathrm{d}\theta$ **1**

$= -\frac{1}{4}\Big[\tan\theta\ \Big]_{\frac{\pi}{3}}^{\frac{\pi}{4}}$ Learn the trig ratios of special angles, such as $\frac{\pi}{4}$ and $\frac{\pi}{3}$ so you can leave your answer in surd form as requested.

$= -\frac{1}{4}\left(1 - \sqrt{3}\right)$

$= \underline{\frac{1}{4}\left(\sqrt{3} - 1\right)}$ **2**

8 NUMERICAL METHODS

Answer	Mark	Examiner's tip

1 Using $x_{n+1} = 2(1+e^{-x_n})$, $x_1 = 0$ gives

$x_2 = 2(1+e^0) = 4$ **1**   Show sufficient working, for the first few iterations, to illustrate use of the iteration formula.

$x_3 = 2(1+e^{-4}) = 2.03663...$

Continuing in the same way gives

$x_4 = 2.260\ 93...$ List the results of the later iterations without showing all the working.

$x_5 = 2.208\ 50...$

$x_6 = 2.219\ 72...$ A graphic calculator is very well suited to this process. e.g. Try

$x_7 = 2.217\ 27...$ $0 \to X$

$x_8 = 2.217\ 81...$ $2(1 + e^{-X}) \to X$

$x_9 = 2.217\ 69...$

$x_{10} = 2.217\ 71...$ **1**   Continue until you are confident that there there will be no further changes that will affect the third decimal place.

$x_{11} = 2.217\ 71...$ **1**

$\therefore \underline{\alpha = 2.218}$ correct to 3 decimal places.   **1**

An equation of which α is a root is
$\underline{x = 2(1 + e^{-x})}$ **1**

2 (a) $x^3 + 3x^2 - 7 = 0 \Rightarrow x^2(x+3) - 7 = 0$ The fact that the rearranged form involves taking the $\sqrt{}$ at the final stage suggests that we need to isolate a quadratic factor.

$\Rightarrow x^2 = \dfrac{7}{x+3} \Rightarrow x = \sqrt{\dfrac{7}{x+3}}$

(assuming $x > 0$)   **1**

Hence $\underline{a = 7}$ and $\underline{b = 3}$ **1**

(b) Taking $x_0 = 2$,

$x_1 = \sqrt{\dfrac{7}{5}} = 1.18321....$ **1**   Note:
In this case, the graph of
$y = x^3 + 3x^2 - 7$ shows that the only real solution of

Continuing in the same way, $x^3 + 3x^2 - 7 = 0$ is positive. In

$x_2 = 1.293\ 58...$ **1**   cases where negative solutions exist it is necessary to consider a

$x_3 = 1.276\ 84...$ rearrangement of the form

$x_4 = 1.279\ 34...$ **1** $x = -\sqrt{g(x)}$.

The approximate solution given by x_4
is $\underline{1.28}$. Comparing the values of   **1**
x_2, x_3 and x_4 it appears that the the third
decimal place is settling to a figure which
will round off to 8.   **1**

Answer	Mark	Examiner's tip

3 Let $f(x) = x^3 - x^2 - 2$

then $f(1) = -2 < 0$ and $f(2) = 2 > 0$ **2**

and so $f(x) = 0$ has a root α between 1 and 2.

It simply needs to be established that there is a change of sign in the value of $f(x)$.

(a) Using $x_{n+1} = x_n - \dfrac{f(x_n)}{f'(x_n)}$ gives

$$x_{n+1} = x_n - \frac{x_n^{\,3} - x_n^{\,2} - 2}{3x_n^{\,2} - 2x_n}$$ **2**

Taking $x_1 = 1.5$ gives

$$x_2 = 1.5 - \frac{1.5^3 - 1.5^2 - 2}{3 \times 1.5^2 - 2 \times 1.5}$$ **1**

so the second approximation is
<u>1.733</u> (to 3 d.p.)   **1**

(b) $x^3 - x^2 - 2 = 0 \Rightarrow x^3 = x^2 + 2$

$\Rightarrow x = \sqrt[3]{(x^2 + 2)}$ **2**

Taking $x_{n+1} = \sqrt[3]{(x_n^{\,2} + 2)}$, $x_1 = 1.5$ gives   **1**

$x_2 = \sqrt[3]{(1.5^2 + 2)} = \underline{1.620}$ (to 3 d.p.)   **1**

$x_3 = \underline{1.666}$ (to 3 d.p.)   **1**

The values of the approximations found in parts (a) and (b) are different simply because the process has not been continued long enough for the results to converge. After several more iterations both processes converge to 1.695 620 8 to seven decimal places.

4 (a) $f(x) = e^x - 5x \Rightarrow f'(x) = e^x - 5$ **3**

(b) $f'(x) = 0 \Rightarrow e^x = 5 \Rightarrow x = \ln 5$ **2**

giving $\underline{x = 1.61}$ to 2 decimal places.   **1**

(c) $f(0.2) = 0.221... > 0$, $f(0.3) = -0.150... < 0$.   **1**

The change of sign shows that there is a root α of $f(x) = 0$ such that $0.2 < \alpha < 0.3$.   **1**

(d) By inspection

$f(2.5) = -0.317... < 0$

$f(2.6) = 0.463... > 0$ **2**

and so $2.5 < \beta < 2.6$

$\therefore \underline{p = 25}$ **2**

Given that p is an integer it follows that $\dfrac{p}{10}$ contains one decimal place. This suggests that we should, first, determine β to 1 d.p. and then interpret the result in terms of p.

It is perfectly valid to locate β by drawing the graph of $y = f(x)$, using a graphics calculator, and then using the Trace function.

Answer	Mark	Examiner's tip

5

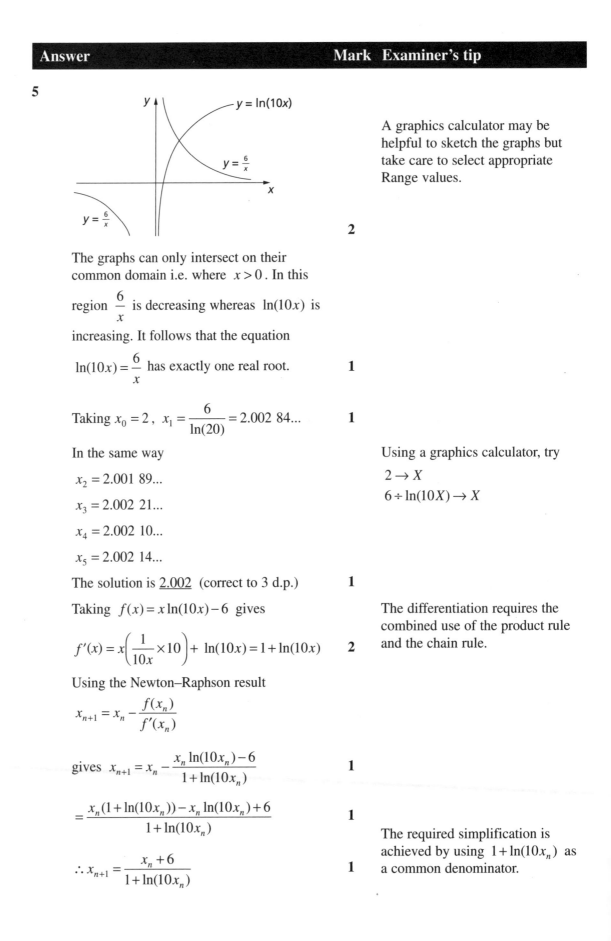

A graphics calculator may be helpful to sketch the graphs but take care to select appropriate Range values.

2

The graphs can only intersect on their common domain i.e. where $x > 0$. In this region $\dfrac{6}{x}$ is decreasing whereas $\ln(10x)$ is increasing. It follows that the equation $\ln(10x) = \dfrac{6}{x}$ has exactly one real root. **1**

Taking $x_0 = 2$, $x_1 = \dfrac{6}{\ln(20)} = 2.002\,84...$ **1**

In the same way

$x_2 = 2.001\,89...$

$x_3 = 2.002\,21...$

$x_4 = 2.002\,10...$

$x_5 = 2.002\,14...$

Using a graphics calculator, try

$2 \rightarrow X$

$6 \div \ln(10X) \rightarrow X$

The solution is <u>2.002</u> (correct to 3 d.p.) **1**

Taking $f(x) = x\ln(10x) - 6$ gives

$f'(x) = x\left(\dfrac{1}{10x} \times 10\right) + \ln(10x) = 1 + \ln(10x)$ **2**

The differentiation requires the combined use of the product rule and the chain rule.

Using the Newton–Raphson result

$x_{n+1} = x_n - \dfrac{f(x_n)}{f'(x_n)}$

gives $x_{n+1} = x_n - \dfrac{x_n \ln(10x_n) - 6}{1 + \ln(10x_n)}$ **1**

$= \dfrac{x_n(1 + \ln(10x_n)) - x_n \ln(10x_n) + 6}{1 + \ln(10x_n)}$ **1**

$\therefore x_{n+1} = \dfrac{x_n + 6}{1 + \ln(10x_n)}$ **1**

The required simplification is achieved by using $1 + \ln(10x_n)$ as a common denominator.

Answer	Mark	Examiner's tip

6 (a) Shaded area $= \dfrac{r^2\theta}{2} - \dfrac{r^2\sin\theta}{2} = \dfrac{r^2}{2}(\theta - \sin\theta)$ **3**   Since θ is in radians, the area of the sector is $\dfrac{r^2\theta}{2}$.

(b) From given result

$$\frac{r^2}{2}(\theta - \sin\theta) = \frac{\pi r^2}{6} \Rightarrow \theta - \sin\theta = \frac{\pi}{3}$$

$$\underline{\sin\theta = \theta - \frac{\pi}{3}}$$ **1**

(c)

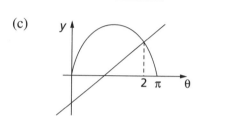

The point of intersection is close to 2 which verifies that $\theta = 2$ is an approximate solution of $\sin\theta = \theta - \dfrac{\pi}{3}$ **1**

1   The sketch needs to be sufficiently accurate, and the horizontal scale sufficiently detailed, to verify the given result.

If you use a calculator make sure that it is set to work in radians.

(d) $\sin\theta = \theta - \dfrac{\pi}{3} \Rightarrow \sin\theta - \theta + \dfrac{\pi}{3} = 0$

Let $f(\theta) = \sin\theta - \theta + \dfrac{\pi}{3}$ then $f'(\theta) = \cos\theta - 1$ **1**

Using Newton's rule a better approximation

is given by $2 - \dfrac{\sin(2) - 2 + \dfrac{\pi}{3}}{\cos(2) - 1} = \underline{1.97\ (2\ \text{d.p.})}$ **2**

9 MATHEMATICS OF UNCERTAINTY

Answer	Mark	Examiner's tip

1 (a)

Weight of fish (x lb)	Class width	Frequency	Frequency density
$0 \le x < 1$	1	21	21
$1 \le x < 1.5$	0.5	32	64
$1.5 \le x < 2.0$	0.5	33	66
$2.0 \le x < 2.5$	0.5	24	48
$2.5 \le x < 3$	0.5	18	36
$3 \le x < 4$	1	21	21
$4 \le x < 5$	1	16	16
$5 \le x < 6$	1	12	12
$6 \le x < 8$	2	11	5.5
$8 \le x < 13$	5	12	2.4

2

You cannot just take the height of the bar to be the frequency, since the intervals are not equal.

Take time to work out the correct boundary points and frequency densities, where

$$\text{frequency density} = \frac{\text{frequency}}{\text{class width}}$$

Writing out the table helps you to clarify your thoughts.

Answer	Mark	Examiner's tip

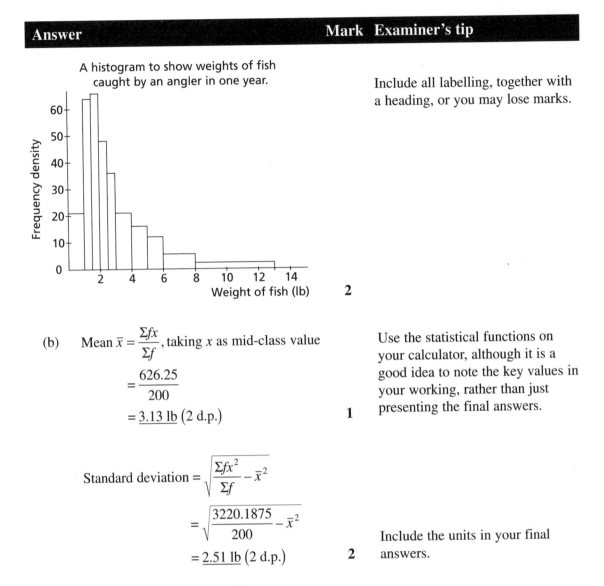

Include all labelling, together with a heading, or you may lose marks.

2

(b) Mean $\bar{x} = \dfrac{\Sigma fx}{\Sigma f}$, taking x as mid-class value

$$= \frac{626.25}{200}$$

$$= \underline{3.13 \text{ lb}}\ (2 \text{ d.p.})$$

1

Use the statistical functions on your calculator, although it is a good idea to note the key values in your working, rather than just presenting the final answers.

Standard deviation $= \sqrt{\dfrac{\Sigma fx^2}{\Sigma f} - \bar{x}^2}$

$$= \sqrt{\frac{3220.1875}{200} - \bar{x}^2}$$

$$= \underline{2.51 \text{ lb}}\ (2 \text{ d.p.})$$

2

Include the units in your final answers.

2 (a) (i) Cumulative frequency table

Speed	≤10	≤20	≤25	≤30	≤35	≤40	≤50	≤70
Cumulative frequency	0	18	44	134	192	224	242	250

2

A common mistake is to plot the cumulative frequencies against mid-points. Remember that upper class boundaries are needed.

The points can be joined with straight lines to form a cumulative frequency polygon, or with a curve.

Label the axes and give a heading.

3

Answer	Mark	Examiner's tip

(ii) $n = 250$, so median
$$= 125^{\text{th}} \text{ value} = \underline{29.5 \text{ mph}}$$
lower quartile
$$= 62.5^{\text{th}} \text{ value} = \underline{26 \text{ mph}}$$
upper quartile
$$= 187.5^{\text{th}} \text{ value} = \underline{34.5 \text{ mph}} \qquad \textbf{3}$$

There will be slight variations in these values, depending on how you joined your points. It is advisable to show your working on your graph.

(b) (i) The extremes of the whiskers are at 10 and 70.

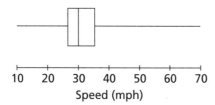

The box plot can be drawn horizontally or vertically. You could place it underneath your curve, using the same horizontal scale.

2

(ii) About half the vehicles break the speed limit, with one quarter breaking it by more than 5 mph. The whisker extends to 70 mph, indicating that some drivers are travelling at over twice the legal speed.

2

Try to make at least two points here.

3 (i) (a) It would be difficult to draw a histogram because it could not convey accurately such vastly different interval widths and column heights.

2

(b) The last class is open-ended and although there are not many share-holders in this category, the size of the shareholding is so large that the total from it would make a significant contribution to the calculation of the mean.

1

Two marks have been allocated, so try to pick out at least two points. You could think of other representations such as a pie chart!

(ii) The median is the $\frac{1}{2}(636\,990)^{\text{th}}$ reading, i.e. the $318\,495^{\text{th}}$ reading, which lies in the interval 100–499.

1

Taking the boundary points as 100 and 500 and noting that $318\,495 - 133\,853 = 184\,642$

the median is $\dfrac{184\,642}{347\,495}$ along this interval

$$\text{median} = 100 + \left(\frac{184\,642}{347\,495}\right) \times 400$$

$$= 312.5\ldots$$

So median size of shareholding is

approximately £310.

2

With such large numbers, it is essential to keep a clear head when deciding which numbers are required. You could use a graphical method, drawing the start of a cumulative frequency curve.

Explain your reasoning.

Answers to Unit 9

Answer	Mark	Examiner's tip

4 $p(A) = 0.4$, $p(B) = 0.7$, $p(A \text{ or } B) = 0.8$

(a) $p(A \text{ or } B) = p(A) + p(B) - p(A \text{ and } B)$

$0.8 = 0.4 + 0.7 - p(A \text{ and } B)$

$\Rightarrow \underline{p(A \text{ and } B) = 0.3}$ — **2**

This is a straightforward application of the probability laws.

(b) $p(A|B) = \dfrac{p(A \text{ and } B)}{p(B)} = \dfrac{0.3}{0.7} = \frac{3}{7}$ — **2**

5

$P(V \text{ and } P) = 0.003 \times 0.95$ $= 0.00285$ — **1**

The easiest way to illustrate the given information is to draw a tree diagram and fill in the missing probabilities.

$P(\bar{V} \text{ and } P) = 0.997 \times 0.02$ $= 0.01994$ — **1**

Note that the 4 extra marks for this question are allocated as follows:
1 using tree
1 first set of branches
1 second set of branches
1 multiplying the probabilities.

$P(\text{Positive result}) = 0.002\ 85 + 0.019\ 94$

$= 0.022\ 79$ — **4**

$P(V|P) = \dfrac{P(V \text{ and } P)}{P(P)} = \dfrac{0.002\ 85}{0.022\ 79}$

$= \underline{0.125}$ (3 s.f.) — **2**

It is helpful if you indicate on your diagram where the values come from.

6 (i)

$* P(F \text{ and } 6) = \frac{1}{2} \times \frac{1}{6} = \frac{1}{12}$

$* P(B \text{ and } 6) = \frac{1}{2} \times \frac{1}{3} = \frac{1}{6}$

To find the conditional probability that the die is biased, given a six is thrown, the sample space has been reduced to the outcomes marked * on the tree.

(a) $P(6) = P(F \text{ and } 6) + P(B \text{ and } 6)$

$= \frac{1}{12} + \frac{1}{6} = \frac{1}{4}$ — **2**

Explain your working.

(b) $P(B|6) = P\dfrac{(B \text{ and } 6)}{P(6)} = \dfrac{\frac{1}{6}}{\frac{1}{4}} = \frac{2}{3}$ — **2**

This is a common type of question.

Answer	Mark	Examiner's tip

(ii) (a) $P(\text{two sixes}) = P(F \text{ and two sixes})$

$\qquad\qquad\qquad + P(B \text{ and two sixes})$

$\qquad\qquad = \frac{1}{2} \times \frac{1}{6} \times \frac{1}{6} + \frac{1}{2} \times \frac{1}{3} \times \frac{1}{3}$

$\qquad\qquad = \frac{1}{72} + \frac{1}{18}$

$\qquad\qquad = \frac{5}{72}$ **3**

In effect, the tree has been extended, but it is easy to imagine which branches are needed.

(b) $P(B \mid \text{two sixes}) = \dfrac{P(B \text{ and two sixes})}{P(\text{two sixes})}$

$\qquad\qquad\qquad = \dfrac{\frac{1}{18}}{\frac{5}{72}} = \frac{4}{5}$ **2**

(iii) $P(B \mid n \text{ sixes})$

$\quad = \dfrac{P(B \text{ and } n \text{ sixes})}{P(n \text{ sixes})}$ **1**

$\quad = \dfrac{\frac{1}{2} \times \left(\frac{1}{3}\right)^n}{\frac{1}{2} \times \left(\frac{1}{6}\right)^n + \frac{1}{2} \times \left(\frac{1}{3}\right)^n}$ Cancel $\frac{1}{2}$,   **1**

$\quad = \dfrac{\left(\frac{2}{6}\right)^n}{\left(\frac{1}{6}\right)^n + \left(\frac{2}{6}\right)^n}$ write $\frac{1}{3}$ as $\frac{2}{6}$,

$\quad = \dfrac{\frac{2^n}{6^n}}{\frac{1}{6^n} + \frac{2^n}{6^n}}$

$\quad = \dfrac{2^n}{1 + 2^n}$ multiply by 6^n **2**

You are being asked to generalise. Look for clues in the patterns and method already established.

In simplifying the expression, you are using the theory of equivalent fractions.

10 VECTORS

Answer	Mark	Examiner's tip

1   $\mathbf{a} = 2\mathbf{i} - \mathbf{k}, \quad \mathbf{b} = \mathbf{i} + 2\mathbf{j} + \mathbf{k}, \quad \mathbf{c} = -\mathbf{j} + \mathbf{k}$

(a) $\mathbf{a}.\mathbf{b} + \mathbf{a}.\mathbf{c} = (2 + 0 - 1) + (0 + 0 - 1) = \underline{0}$ **3**

(b) $\mathbf{a}.\mathbf{b} + \mathbf{a}.\mathbf{c} = \mathbf{a}.(\mathbf{b} + \mathbf{c})$ **1**

$\quad \therefore \mathbf{a}.(\mathbf{b} + \mathbf{c}) = 0$

$\quad$ Since $|\mathbf{a}| \neq 0$ and $|\mathbf{b} + \mathbf{c}| \neq 0$ it follows that

$\quad \underline{\mathbf{b} + \mathbf{c} \text{ is perpendicular to } \mathbf{a}}.$ **1**

The scalar product is distributive over vector addition. In other words, the common factor $\mathbf{a}$ can be taken outside the brackets in the usual way.

Answer	Mark	Examiner's tip
2 (i) $\overrightarrow{ON} = \overrightarrow{OC} + \overrightarrow{CG} + \overrightarrow{GN}$	1	Show sufficient working to make your method clear.
$= 2\mathbf{j} + 2\mathbf{k} + \mathbf{i}$		Note that any 'route' which starts at O and finishes at N will be equivalent in vector terms.
$= \mathbf{i} + 2\mathbf{j} + 2\mathbf{k}$	1	
$\overrightarrow{MG} = \overrightarrow{MB} + \overrightarrow{BF} + \overrightarrow{FG}$		
$= \mathbf{j} + 2\mathbf{k} - 2\mathbf{i}$		
$= -2\mathbf{i} + \mathbf{j} + 2\mathbf{k}$	1	
(ii) $\overrightarrow{ON}.\overrightarrow{MG} = (\mathbf{i} + 2\mathbf{j} + 2\mathbf{k}).(-2\mathbf{i} + \mathbf{j} + 2\mathbf{k})$		
$= -2 + 2 + 4 = 4$	1	
but $\overrightarrow{ON}.\overrightarrow{MG} = ON \times MG \cos\theta$	1	Note the distinction between, for example, the vector $\overrightarrow{ON}$ and the distance (scalar) ON.
$= \sqrt{1^2 + 2^2 + 2^2} \times \sqrt{(-2)^2 + 1^2 + 2^2} \times \cos\theta$		
$= 9\cos\theta$	1	
so $9\cos\theta = 4$		Alternatively, you could use the formula $\theta = \cos^{-1}\left(\dfrac{\mathbf{a}.\mathbf{b}}{\|\mathbf{a}\|\|\mathbf{b}\|}\right)$
giving $\theta = \cos^{-1}\left(\frac{4}{9}\right) = 63.612...°$	1	
and so the acute angle between the directions of $\overrightarrow{ON}$ and $\overrightarrow{MG}$ is $63.6°$,		Interpret the calculated value of θ in terms of the original problem.
correct to the nearest $0.1°$.	1	
3 (a) $\mathbf{r}$ and $\mathbf{s}$ are perpendicular $\Rightarrow \mathbf{r}.\mathbf{s} = 0$	1	Make a clear statement and substitute the given information.
$(\lambda\mathbf{i} + (2\lambda - 1)\mathbf{j} - \mathbf{k}).((1 - \lambda)\mathbf{i} + 3\lambda\mathbf{j} +$		
$(4\lambda - 1)\mathbf{k}) = 0$	1	
$\lambda(1 - \lambda) + 3\lambda(2\lambda - 1) - (4\lambda - 1) = 0$	2	Use brackets and take care with the negatives.
$\lambda - \lambda^2 + 6\lambda^2 - 3\lambda - 4\lambda + 1 = 0$		
$5\lambda^2 - 6\lambda + 1 = 0$	1	The fact that the quadratic factorises suggests that the working is correct but, time permitting, the result could easily be checked by substituting for λ in $\mathbf{r}$ and $\mathbf{s}$.
$(5\lambda - 1)(\lambda - 1) = 0$		
The values of λ for which $\mathbf{r}$ and $\mathbf{s}$ are perpendicular are <u>0.2 and 1</u>	2	
(b) substituting $\lambda = 2$ gives		
$\overrightarrow{OA} = 2\mathbf{i} + 3\mathbf{j} - \mathbf{k}$, $\overrightarrow{OB} = -\mathbf{i} + 6\mathbf{j} + 7\mathbf{k}$	1	
$\overrightarrow{AB} = \overrightarrow{OB} - \overrightarrow{OA}$		
$= -\mathbf{i} + 6\mathbf{j} + 7\mathbf{k} - (2\mathbf{i} + 3\mathbf{j} - \mathbf{k})$		
$= -3\mathbf{i} + 3\mathbf{j} + 8\mathbf{k}$	1	

Letts
Q&A

Answer	Mark	Examiner's tip

(c)

Angle *BAO* is the angle between the vectors $\overrightarrow{AB}$ and $\overrightarrow{AO}$.

1 — An important statement.

$$\overrightarrow{AB}.\overrightarrow{AO} = (-3\mathbf{i}+3\mathbf{j}+8\mathbf{k}).(-2\mathbf{i}-3\mathbf{j}+\mathbf{k})$$

$$= 6 - 9 + 8 = 5$$

1

but

$$\overrightarrow{AB}.\overrightarrow{AO} = AB \times AO \cos\theta$$

Make your method clear by showing all of the necessary working.

$$= \sqrt{(-3)^2 + 3^2 + 8^2} \times \sqrt{(-2)^2 + (-3)^2 + 1^2} \times \cos\theta \qquad \textbf{1}$$

$$= \sqrt{82} \times \sqrt{14} \cos\theta$$

and so $\cos\theta = \dfrac{5}{\sqrt{82 \times 14}}$ **1**

giving $\theta = \cos^{-1}\left(\dfrac{5}{\sqrt{82 \times 14}}\right) = 81.51...°$

hence, <u>angle *BAO* = 82°</u> to the nearest degree. **1**

Give the final answer to the required level of accuracy.

4 (a) (i) For the lines to intersect, there must be some value of *s* and *t* such that

$$2\mathbf{i} + s(\mathbf{i}+3\mathbf{j}+4\mathbf{k}) = \mathbf{k} + t(\mathbf{i}+\mathbf{j}+\mathbf{k}) \qquad \textbf{1}$$

Equating corresponding components gives

$2 + s = t$	(1)	
$3s = t$	(2)	**1**
$4s = t + 1$	(3)	

The single vector equation gives three separate equations in the two parameters *s* and *t*. All three equations must be stated.

From (1) and (2) $\quad 3s = 2 + s \Rightarrow s = 1$

Substituting for *s* in (1) gives $t = 3$ **1**

Checking for consistency in (3) gives

L.H.S. = $4 \times 1 = 4$, R.H.S. = $3 + 1 = 4$.

Since L.H.S. = R.H.S it follows that the equations are consistent and so the lines intersect. **1**

Substituting $s = 1$ in the equation for *l* gives the position vector of the point of intersection as

$$\underline{3\mathbf{i} + 3\mathbf{j} + 4\mathbf{k}} \qquad \textbf{1}$$

A value for *s* and *t* may be found using any pair of equations.

To establish that the lines intersect, the values of *s* and *t* must be substituted in the remaining equation in order to check that the three equations are **consistent**. Avoid presenting a chain of reasoning which concludes that $4 = 4$.

Letts
Q&A

Answer	Mark	Examiner's tip

(ii) The angle between the lines is given by the angle betwen the vectors
$\mathbf{i}+3\mathbf{j}+4\mathbf{k}$ and $\mathbf{i}+\mathbf{j}+\mathbf{k}$.   **1**   These are the direction vectors of the two lines.

$(\mathbf{i}+3\mathbf{j}+4\mathbf{k}).(\mathbf{i}+\mathbf{j}+\mathbf{k})=1+3+4=8$ **1**

giving

$\sqrt{1^2+3^2+4^2}\times\sqrt{1^2+1^2+1^2}\times\cos\theta=8$

$\cos\theta=\dfrac{8}{\sqrt{26\times3}}$

$\theta=\cos^{-1}\left(\dfrac{8}{\sqrt{78}}\right)=25.06..°$ **1**

$\therefore$ <u>acute angle between lines = 25°</u>   **1**
to the nearest degree.

(b) At the points on l which are $5\sqrt{10}$ units

from the origin we have $|\mathbf{r}|=5\sqrt{10}$ **1**   Interpret the given condition as an equation to solve.

giving $\sqrt{(s+2)^2+9s^2+16s^2}=5\sqrt{10}$ **1**

$s^2+4s+4+9s^2+16s^2=250$

$26s^2+4s-246=0$

$13s^2+2s-123=0$

$(13s+41)(s-3)=0$

$s=3$ or $s=-\frac{41}{13}$ **1**

Hence, the required points have position vectors $\underline{2\mathbf{i}+3(\mathbf{i}+3\mathbf{j}+4\mathbf{k})}$ and

$\underline{2\mathbf{i}-\frac{41}{13}(\mathbf{i}+3\mathbf{j}+4\mathbf{k})}$.   **2**

(c) At the closest point   The closest point has the property

$(2\mathbf{i}+s(\mathbf{i}+3\mathbf{j}+4\mathbf{k})-(6\mathbf{i}-\mathbf{j}+3\mathbf{k})).(\mathbf{i}+3\mathbf{j}+4\mathbf{k})=0$ **2**   that its direction from the given point is perpendicular to the

$(\mathbf{i}(s-4)+\mathbf{j}(3s+1)+\mathbf{k}(4s-3)).(\mathbf{i}+3\mathbf{j}+4\mathbf{k})=0$ **1**   direction of the line.

$s-4+3(3s+1)+4(4s-3)=0$ **1**

$26s-13=0$ and so $s=0.5$ **1**

$\therefore$ The required position vector is

$\underline{2\mathbf{i}+0.5(\mathbf{i}+3\mathbf{j}+4\mathbf{k})}$ **1**

5 (a) (i) $\overrightarrow{OD}=\dfrac{\mathbf{b}+2\mathbf{c}}{3}$ $\overrightarrow{OE}=\dfrac{3\mathbf{a}+\mathbf{c}}{4}$ **1**   These results are based on the formula for the position vector of a point dividing a line in a given ratio.

Answer	Mark	Examiner's tip

(ii) $\overrightarrow{AD} = \overrightarrow{OD} - \overrightarrow{OA} = \dfrac{\mathbf{b} + 2\mathbf{c}}{3} - \mathbf{a}$

$\qquad\qquad = \dfrac{\mathbf{b} + 2\mathbf{c} - 3\mathbf{a}}{3}$ **1**

The position vector of any point on AD is given by $\mathbf{r} = \mathbf{a} + s(\mathbf{b} + 2\mathbf{c} - 3\mathbf{a})$ **1**

When $s = \frac{1}{9}$, $\mathbf{r} = \frac{2}{3}\mathbf{a} + \frac{1}{9}\mathbf{b} + \frac{2}{9}\mathbf{c}$ which   **1**

therefore is the position vector of a point on AD.

Note that the direction vector has been taken as $\mathbf{b} + 2\mathbf{c} - 3\mathbf{a}$ rather than $\dfrac{\mathbf{b} + 2\mathbf{c} - 3\mathbf{a}}{3}$.

$\overrightarrow{BE} = \overrightarrow{OE} - \overrightarrow{OB} = \dfrac{3\mathbf{a} + \mathbf{c}}{4} - \mathbf{b}$

$\qquad\qquad = \dfrac{3\mathbf{a} - 4\mathbf{b} + \mathbf{c}}{4}$

The value $s = \frac{1}{9}$ is found by inspection.

The position vector of any point on BE is given by $\mathbf{r} = \mathbf{b} + t(3\mathbf{a} - 4\mathbf{b} + \mathbf{c})$ **1**

When $t = \frac{2}{9}$, $\mathbf{r} = \frac{2}{3}\mathbf{a} + \frac{1}{9}\mathbf{b} + \frac{2}{9}\mathbf{c}$ which

therefore is the position vector of a point on BE.   **1**

Hence, the position vector of G is $\frac{2}{3}\mathbf{a} + \frac{1}{9}\mathbf{b} + \frac{2}{9}\mathbf{c}$.   **1**

It has now been established that the given position vector lies on both lines and so represents the position of G.

(iii) $\overrightarrow{CG} = \frac{2}{3}\mathbf{a} + \frac{1}{9}\mathbf{b} + \frac{2}{9}\mathbf{c} - \mathbf{c} = \frac{2}{3}\mathbf{a} + \frac{1}{9}\mathbf{b} - \frac{7}{9}\mathbf{c}$

$\overrightarrow{OF} = \mathbf{c} + \mu(6\mathbf{a} + \mathbf{b} - 7\mathbf{c})$ **1**

$\qquad = 6\mu\mathbf{a} + \mu\mathbf{b} + (1 - 7\mu)\mathbf{c}$

Direct comparison of the two results for $\overrightarrow{OF}$ is made easier by expressing both in the same form.

F also lies on AB so

$\overrightarrow{OF} = \mathbf{b} + \lambda(\mathbf{a} - \mathbf{b}) = \lambda\mathbf{a} + (1 - \lambda)\mathbf{b}$ **1**

by inspection, $\mu = \frac{1}{7}$, $\lambda = \frac{6}{7}$.

Hence $\underline{\overrightarrow{OF} = \frac{6}{7}\mathbf{a} + \frac{1}{7}\mathbf{b}}$ **1**

(b) (i)

The diagram helps to make the relationships clear.

$\overrightarrow{OQ} = \mathbf{p} + \mathbf{r} \qquad \overrightarrow{PR} = \mathbf{r} - \mathbf{p}$ **2**

(ii) $\overrightarrow{OQ} \cdot \overrightarrow{PR} = (\mathbf{p} + \mathbf{r}) \cdot (\mathbf{r} - \mathbf{p})$ **1**

$\qquad\qquad = \mathbf{r} \cdot \mathbf{r} - \mathbf{p} \cdot \mathbf{p}$

$\qquad\qquad = OR^2 - OP^2$ **1**

$\qquad\qquad = 0$ (since $OR = OP$).

So OQ and PR are perpendicular.   **1**

11 MECHANICS

Answer	Mark	Examiner's tip

1 Using $s = ut + \frac{1}{2}at^2$ gives $2 = \frac{1}{2} \times 9.8t^2$ ($u = 0$) — **2**

and so $t = \sqrt{\dfrac{4}{9.8}} = 0.6388...$

∴ Time taken = <u>0.64 s</u> (2 d.p.) — **1**

Examiner's tip: This is a straightforward question that simply requires selection of the appropriate formula. The value of g was given as 9.8 ms^{-2} in the general rubric for the paper.

2

Horizontal motion (assumed constant)

Horizontal component = $20\cos 60 = \underline{10 \text{ ms}^{-1}}$ — **1**

Vertical motion (when $t = 2$)

Using $v = u + at$ gives $v = 20\sin 60 - 9.8 \times 2$

∴ $v = -2.279...$ — **2**

∴ Vertical component = $\underline{2.3 \text{ ms}^{-1}}$ downwards — **1**

Examiner's tip: It is always good practice to draw a simple diagram, labelled with the relevant information. Be clear about which vertical direction is to be regarded as *positive*.

Taking upwards as positive $a = -g$.

The negative sign of v is significant.

3 (i) Horizontal component = $\underline{19.1 \text{ ms}^{-1}}$ — **1**

Vertical component = $\underline{11 \text{ ms}^{-1}}$ — **1**

(ii) (Using $s = ut + \frac{1}{2}at^2$ in the vertical direction, the required result follows immediately). — **2**

The ball is at least 2 m above the ground for $\underline{0.2 \le t \le 2}$ — **3**

(iii) (Taking $t = 0.2$ and using the result for the horizontal component established in (i), the result follows.)

(Using the fact that the ball is above the required height for between 0.2 and 2 seconds and, again using the result for the horizontal component suggests...) <u>19 vans</u>. — **7**

Examiner's tip: Draw your own diagram to check these results. Use of a diagram can help you spot obvious errors such as having your calculator set for radians instead of degrees!

Set $y = 2$ and solve the inequality. A sketch graph is helpful.

Pay particular attention to the structure of the question in order to make use of the results established in the early parts.

4 (a) (i) $x = V\cos\theta . t$ (1) — **2**

$y = V\sin\theta . t - \frac{1}{2}gt^2$.   (2)

Answer	Mark	Examiner's tip

(ii) From (1), $t = \dfrac{x}{V\cos\theta}$.

Substituting for t in (2) gives

$$y = V\sin\theta . \frac{x}{V\cos\theta} - \frac{g}{2} . \frac{x^2}{V^2\cos^2\theta} \qquad 1$$

$$= x\tan\theta - \left(\frac{gx^2}{2V^2}\right) . \frac{1}{\cos^2\theta}$$

Hence $y = x\tan\theta - \left(\dfrac{gx^2}{2V^2}\right)\sec^2\theta$ 1

Equations (1) and (2) are essentially parametric equations and the required result is the cartesian equation of the flight path. The result follows by eliminating t from the equations.

(b)

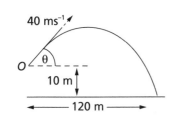

A diagram can help to clarify the nature of the problem.

$$-10 = 120\tan\theta - \left(\frac{10\times120^2}{2\times40^2}\right)\sec^2\theta \qquad 2$$

$$\Rightarrow -10 = 120\tan\theta - 45\sec^2\theta$$

$$\Rightarrow 45(1+\tan^2\theta) - 120\tan\theta - 10 = 0 \qquad 1$$

$$\Rightarrow 9\tan^2\theta - 24\tan\theta + 7 = 0$$

$$\Rightarrow (3\tan\theta - 1)(3\tan\theta - 7) = 0 \qquad 1$$

giving $\theta = \tan^{-1}\left(\frac{1}{3}\right)$ or $\theta = \tan^{-1}\left(\frac{7}{3}\right)$ 2

Substitute the given information into the trajectory formula. Note $y < 0$.

Use the identity $\sec^2\theta \equiv 1 + \tan^2\theta$ to produce a quadratic in $\tan\theta$.

The given result $\theta = \tan^{-1}\left(\frac{1}{3}\right)$ suggests using $3\tan\theta$ in both brackets when factorising.

5 (a) (i) The diagram shows the *external* forces acting on the system.

1

The forces shown are in equilibrium.
Hence $T = 950g + 50g = 1000g$ 1

The tension in the lift cable is 10 000 N   1

Consider a 'system' comprising the lift and the woman.

The key point here is that the system is moving with uniform speed and so these external forces must be in equilibrium. Don't confuse the situation by labelling internal forces between the elements of the system at this stage.

Answer	Mark	Examiner's tip

(ii) The diagram shows the forces acting on the woman.

R

50g — 1

The forces shown are in equilibrium, hence $R = 50g$. — 1

Force exerted on woman by floor of lift = <u>500 N</u>

The focus of attention has now switched to the forces acting on just one part of the system i.e. those *acting on the woman.* Again, these forces must be in equilibrium. It is *essential* to be clear about the nature of the forces represented in a diagram and on the precise focus of attention.

(b) (i) For the complete system

Resultant upward force = $T - 1000g$ — 1

Using $F = ma$ gives — 2

$T - 1000g = 1000 \times 2$

$\Rightarrow T = 12\,000$ N

$\therefore$ Tension in cable is now <u>12 000 N</u> — 1

At this stage m represents the mass of the whole system.

(ii) In the same way, for the woman

$R - 50g = 50 \times 2 \Rightarrow R = 600$ N — 2

$\therefore$ Force exerted on woman by floor of lift = <u>600 N.</u> — 1

At this point it is the mass of the woman that is relevant.

6 Let the tension in the string be T and let the acceleration of the system be a.

The diagram shows the forces acting on the two bodies.

T   T

2 kg ●   ● 3 kg

2g   3g

For the 2 kg mass $\quad T - 2g = 2a \qquad$ (1) — 1

For the 3 kg mass $\quad 3g - T = 3a \qquad$ (2) — 1

Use $F = ma$ for each body, separately, to produce two simultaneous equations.

(1)+(2) gives $g = 5a \Rightarrow a = \dfrac{g}{5}$

(1)–(2) gives $2T - 5g = -a \Rightarrow T = \dfrac{5g - a}{2}$ — 1

Taking $g = 10$ ms^{-2} gives $a = 2$ ms^{-2} and $T = 24$ N

Hence, the tension in the string is <u>24 N.</u> — 1

Check that your answers are reasonable. A value of $a > g$, for example, would be impossible.

(The assumptions included in the model are: The string is light and inextensible, the pulley is light and frictionless and the parts of the system are not involved in any collisions.) — 2

Only two assumptions need to be stated – one mark each.

Answer	Mark	Examiner's tip

7

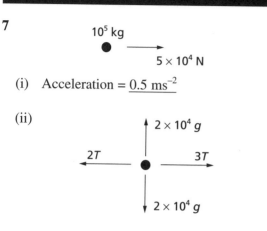

(i) Acceleration = <u>0.5 ms^{-2}</u> **4**

(ii)

(iii) $T = 10^4$ N **3**

The tensions in the couplings are:

Rear	10^4 N	**1**
Middle	2×10^4 N	**1**
Front	3×10^4 N	**1**

(iv) Resultant force on train is 4×10^4 N giving a new acceleration of the complete system of 0.4 ms^{-2}. **2**

The tension in the coupling to the last truck $= 8 \times 10^3$ N

The tension in the coupling to the last truck is <u>reduced to 8×10^3 N</u> **2**

In this question the 'system' is made up of the engine and the three trucks. In the given model, the only external force in the direction of motion is 5×10^4 N.

Note: The total mass should be converted to S.I. units (i.e. kg) before using $F = ma$.

The resultant force on each truck must be the same since they each have the same mass and acceleration.

Use $F = ma$, where the resultant force on any truck is given by T, the mass of each truck is 2×10^4 kg and the common acceleration is 0.5 ms^{-2}.

8

(a) The plank should be modelled as a uniform rod supported at its end points. **1**

The man should be modelled as a particle. **1**

(b) For equilibrium

$2F + F = 100g + 140g$ **3**

so $F = 80g$

Force exerted by support at B is <u>784 N</u>. **1**

(c) Taking moments about A

$80g \times 5 = 140g \times 2.5 + 100g \times AC$ **2**

giving $AC = 0.5$ m.

The man is <u>0.5 m</u> from A. **1**

The larger reaction at A suggests that the man is standing nearer to A than to B. The position of C on the diagram doesn't affect any of the calculations but knowing that AC should work out as less than 2.5 m provides a check that the answer is reasonable.

The rubric of the paper gives g as 9.8 ms^{-2}.

The use of moments is required to determine the position of C. For equilibrium, the clockwise and anti-clockwise moments must be equal.

Answers to Unit 11

Answer	Mark	Examiner's tip

9 (a) <u>2 cm</u>   **1**   This result follows from the symmetry of the figure.

(b) The ear-ring might be regarded as the sum of two square laminae of relative masses 4 and –1.   **2**

Since the lamina is uniform, the relative masses of the two squares depends on their relative areas. The use of a *negative* figure to represent a missing element can greatly simplify some calculations.

The corresponding distances of their centres of mass from *AD*, are 2 cm and 3 cm respectively.   **2**

Using $\bar{x} = \dfrac{\sum mx}{\sum m}$ gives $\bar{x} = \dfrac{4 \times 2 - 1 \times 3}{3}$ cm   **1**

Hence G is $\frac{5}{3}$ cm from *AD*.   **1**

(c)

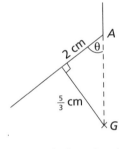

A common feature of many questions relating to centre of mass is that, once its position is found, you may be required to relate this information to an equilibrium position or to conditions for toppling.

Since the figure is hanging freely in equilibrium G must be vertically below A.   **1**

Note that this diagram only includes the key features – don't waste time producing a work of art.

From the diagram $\tan \theta = \frac{5}{6}$ $\therefore \theta = 39.8°$ **2**

So *AD* makes an angle of <u>40°</u> with the downward vertical, to the nearest degree.   **1**

10   Before impact $5u$ u

$P\,(m)$ $Q\,(m)$

After impact v $2v$

A 'before and after' diagram is a useful way of presenting the information, in preparation for applying the conservation of momentum principle.

(a) Momentum is conserved in the collision   **1**   State the principle.

hence $5mu + mu = mv + 2mv$ **2**

$\therefore 6mu = 3mv$

$\therefore v = 2u$ **1**

K.E. before impact

$= \frac{1}{2}m(5u)^2 + \frac{1}{2}mu^2 = 13mu^2$ **1**

K.E. after impact

$= \frac{1}{2}m(2u)^2 + \frac{1}{2}m(4u)^2 = 10mu^2$ **1**

and so the K.E. lost in the collision is <u>$3mu^2$</u>   **1**

Note: The redistribution of momentum corresponds to a loss of K.E.

Answer	Mark	Examiner's tip

(b) Momentum of Q before impact = mu **1**

Momentum of Q after impact = $4mu$ **1**

Hence, magnitude of impulse on Q is <u>$3mu$</u>.   **1**

Note: In the collision, P receives an equal and opposite impulse from Q so that, overall, momentum is conserved.

11

As usual, a simple diagram helps to clarify the nature of the problem.

(a) Gravitational P.E. lost   **1**
= (K.E. + Elastic P.E.) gained

$$mgx = \tfrac{1}{2}mv^2 + \frac{mgx^2}{2a}$$ **2**

$$\therefore 2agx = av^2 + gx^2$$

$$\therefore \underline{av^2 = 2agx - gx^2} \quad (1)$$ **1**

This principle is at the heart of the problem and should be clearly stated.

At maximum distance

$$v = 0 \Rightarrow 2agx - gx^2 = 0$$ **1**

$$\Rightarrow gx(2a - x) = 0 \Rightarrow x = 0 \text{ or } x = 2a.$$

Be aware of the structure of the question. The idea is to relate the given condition to the result (1) already established.

Hence, maximum distance of student below A is <u>$3a$</u>.   **1**

Remember to add the distance from A to O.

(b) The maximum speed occurs when the tension in the rope = the weight of the student.

This gives $\dfrac{mgx}{a} = mg \Rightarrow x = a.$ **1**

Note: Below this point, the tension will be greater than the weight of the student and so the speed will be reduced.

Substituting for x in (1) gives

$$av^2 = 2a^2g - a^2g = a^2g$$

hence $v^2 = ag$ and so the maximum speed

is $\sqrt{ag}$.   **1**

(c) The maximum tension occurs at the greatest distance and is given by

$$\frac{mg \times 2a}{a} = \underline{2mg}$$ **2**

Note: x is measured from O not A.

Answer	Mark	Examiner's tip

12

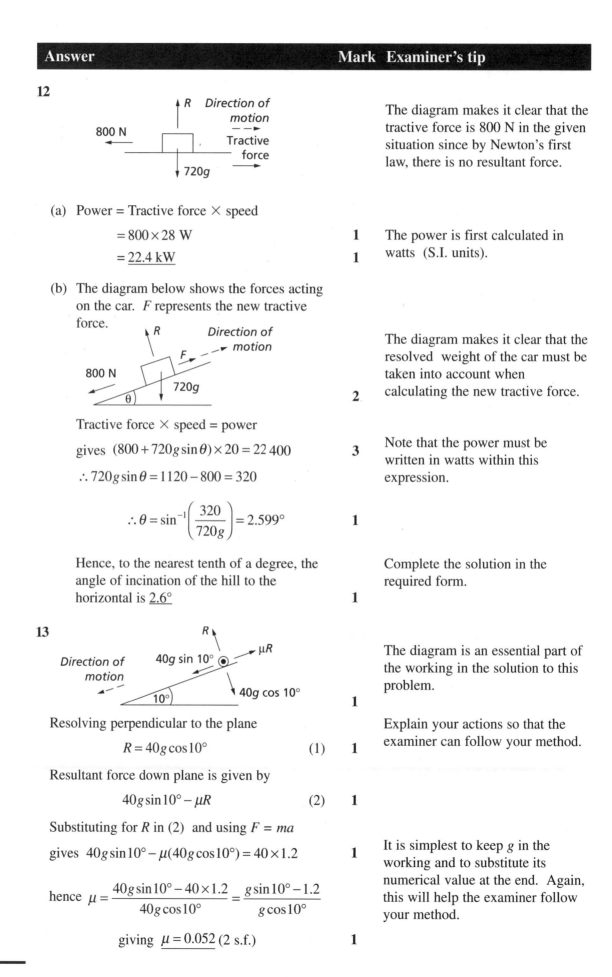

(a) Power = Tractive force × speed

$$= 800 \times 28 \text{ W}$$ — 1

$$= \underline{22.4 \text{ kW}}$$ — 1

The power is first calculated in watts (S.I. units).

The diagram makes it clear that the tractive force is 800 N in the given situation since by Newton's first law, there is no resultant force.

(b) The diagram below shows the forces acting on the car. *F* represents the new tractive force.

The diagram makes it clear that the resolved weight of the car must be taken into account when calculating the new tractive force. — 2

Tractive force × speed = power

gives $(800 + 720g \sin \theta) \times 20 = 22\,400$ — 3

$\therefore 720g \sin \theta = 1120 - 800 = 320$

Note that the power must be written in watts within this expression.

$$\therefore \theta = \sin^{-1}\left(\frac{320}{720g}\right) = 2.599°$$ — 1

Hence, to the nearest tenth of a degree, the angle of incination of the hill to the horizontal is $\underline{2.6°}$ — 1

Complete the solution in the required form.

13

Resolving perpendicular to the plane — 1

$$R = 40g \cos 10° \qquad (1)$$ — 1

The diagram is an essential part of the working in the solution to this problem.

Explain your actions so that the examiner can follow your method.

Resultant force down plane is given by

$$40g \sin 10° - \mu R \qquad (2)$$ — 1

Substituting for *R* in (2) and using $F = ma$

gives $40g \sin 10° - \mu(40g \cos 10°) = 40 \times 1.2$ — 1

hence $\mu = \dfrac{40g \sin 10° - 40 \times 1.2}{40g \cos 10°} = \dfrac{g \sin 10° - 1.2}{g \cos 10°}$

It is simplest to keep *g* in the working and to substitute its numerical value at the end. Again, this will help the examiner follow your method.

giving $\underline{\mu = 0.052}$ (2 s.f.) — 1

Answer	Mark	Examiner's tip

Using $v^2 = u^2 + 2as$, gives $v^2 = 2 \times 1.2 \times 50$ **1**

giving $v^2 = 120 \Rightarrow v = 10.95...$

and so the speed $= \underline{11 \text{ ms}^{-1}}$ (2 s.f.)   **1**

The acceleration is constant and so the formulas for constant acceleration apply.

Using $v = u + at$ gives $10.95... = 1.2t$ **1**

which gives the time taken as $\underline{9.1 \text{ s}}$ (2 s.f.)   **1**

The final answers should be rounded to a level of accuracy appropriate to the conditions of the problem.

Diagram: inclined plane at 10° with forces R, T, μR, 40g cos 10°, 40g sin 10°, and Direction of motion.

The force of friction now acts *down* the plane.

From the diagram

$T = 40g \sin 10° + \mu(40g \cos 10°) = 88.1.. \text{ N}$ **2**

The tension in the rope is $\underline{88 \text{ N}}$ (2 s.f.)   **1**

14 $v = 12t - 3t^2$, $0 \le t \le 5$

The first formula for v applies throughout the time interval for this part of the question.

(a) $\dfrac{ds}{dt} = v \Rightarrow s = \int (12t - 3t^2) \, dt = 6t^2 - t^3 + C$ **2**

when $t = 0$, $s = 0 \Rightarrow C = 0$ **1**

$\therefore s = 6t^2 - t^3$ for values of t such that $0 \le t \le 5$.   **1**

The given approach has used an indefinite integral, in which the constant of integration C has worked out as zero. An alternative approach would have been to

evaluate $\int_0^5 (12t - 3t^2) \, dt$.

Thus, the displacement of P from O when $t = 5$ is given by

$$6 \times 5^2 - 5^3 = \underline{25 \text{ m}}$$ **1**

(b) Displacement when $t = 6$ is given by

$$25 + \int_5^6 -375t^{-2} \, dt = 25 + \left[\frac{375}{t} \right]_5^6$$ **3**

$$= 25 + 62.5 - 75 = \underline{12.5 \text{ m}}$$ **2**

Alternatively, this part could also have been answered using an indefinite integral, giving $s = \dfrac{375}{t} - 50$ (i.e. $C = -50$).

15(i) (a) Using $\mathbf{F} = m\mathbf{a}$ gives

$$4t^2 \mathbf{i} + (2t - 3)\mathbf{j} = 0.2\mathbf{a}$$

$$\Rightarrow \mathbf{a} = 20t^2 \mathbf{i} + (10t - 15)\mathbf{j}$$ **3**

When $t = 2$ the acceleration is $\underline{80\mathbf{i} + 5\mathbf{j}}$ **1**

The formula $\mathbf{F} = m\mathbf{a}$ applies equally well when $\mathbf{F}$ and $\mathbf{a}$ are expressed in vector notation.

(b) $\mathbf{v} = \int \mathbf{a} \, dt = \left(\dfrac{20t^3}{3} + x \right)\mathbf{i} + (5t^2 - 15t + y)\mathbf{j}$ **3**

(ii) $\mathbf{v} = \dfrac{d\mathbf{r}}{dt} = \underline{(2t + 2)\mathbf{i} + (3t^2 - 4)\mathbf{j}}$ **3**

Integrate the components separately and introduce independent constants of integration. Under the given conditions, these constants take the values x and y.

Answer	Mark	Examiner's tip

(iii) Velocity of P relative to Q when $t = 3$ is given by

$$\left(\frac{20 \times 3^3}{3} + x - 8\right)\mathbf{i} + (45 - 45 + y - 27 + 4)\mathbf{j}$$ 2

hence $(172 + x)\mathbf{i} + (y - 23)\mathbf{j} = 180\mathbf{i} + 30\mathbf{j}$

giving $\underline{x = 8}$ and $\underline{y = 53}$ 2

Subtract the velocity of Q from the velocity of P when $t = 3$.

Two separate equations may be formed by comparing the coefficients of $\mathbf{i}$ and the coefficients of $\mathbf{j}$.

16(a) (i) $\mathbf{v} = \dfrac{\mathrm{d}\mathbf{r}}{\mathrm{d}t} = \underline{-2\sin 2t\mathbf{i} + 2\cos 2t\mathbf{j} - 6\mathbf{k}}$ 2

Momentum $= m\mathbf{v}$

$= \underline{-6\sin 2t\mathbf{i} + 6\cos 2t\mathbf{j} - 18\mathbf{k}}$ 1

Differentiate the $\mathbf{i}$, $\mathbf{j}$ and $\mathbf{k}$ components separately. Note the use of the chain rule.

(ii) $\mathbf{F} = m\mathbf{a} = m\dfrac{\mathrm{d}\mathbf{v}}{\mathrm{d}t} = 3(-4\cos 2t\mathbf{i} - 4\sin 2t\mathbf{j})$ 1

$\therefore \mathbf{F} = -12(\cos 2t\mathbf{i} + \sin 2t\mathbf{j})$ 1

(iii) $\mathbf{F}.\mathbf{v} = 24\sin 2t\cos 2t - 24\sin 2t\cos 2t = 0$

hence the direction of $\mathbf{F}$ is perpendicular to the velocity for all values of t.   2

A standard application of the scalar product is to show that two non-zero vector quantities are perpendicular by establishing that their scalar product is zero.

(b) P is moving downwards in a circular spiral at constant speed.   2

17(a) To maintain a circular orbit:
radial acceleration = acceleration due to gravity.   1

This gives $\dfrac{v^2}{r} = 1.6 \Rightarrow \dfrac{v^2}{3 \times 10^6} = 1.6$ 1

$v^2 = 1.6 \times 3 \times 10^6$

$\underline{v = 2190 \text{ ms}^{-1}}$ (to 3 s.f.)   1

Make a clear statement of the principle to be applied to ensure the award of the method mark.

(b) Using $\mathbf{F} = m\mathbf{a}$

gives $\dfrac{Mk}{r^2} = M \times 1.6$ 1

$k = 1.6r^2 = 1.6 \times (2 \times 10^6)^2 \text{ m}^3 \text{ s}^{-2}$ 1

hence $\underline{k = 6.4 \times 10^{12} \text{ m}^3 \text{ s}^{-2}}$ 1

This corresponds to taking the weight of an object on Earth as mg.

Answer	Mark	Examiner's tip
(c) Applying the refined model to the condition for maintaining a circular orbit gives	1	The new model takes into account the variation in the gravitational attraction with distance and so the result should be more accurate.
$$\frac{v^2}{3 \times 10^6} = \frac{6.4 \times 10^{12}}{(3 \times 10^6)^2}$$	1	
$$v = \sqrt{\frac{6.4 \times 10^{12}}{3 \times 10^6}}$$		In some questions you may be asked to explain the consequences of adopting a particular model.
$$\underline{v = 1460 \text{ m s}^{-1} \ (3 \text{ s.f.})}$$	2	

12 STATISTICS

Answer	Mark	Examiner's tip
1 $P(\text{late}) = \frac{49}{126} = \frac{7}{18}$	1	It is *always* helpful to define your variable.
Let X be the number of times the train is late in 5 days.		$p = \frac{7}{18}$ is the best estimate that we have, so it is used in the model.
$X \sim \text{B}(5, \frac{7}{18}) \qquad q = 1 - p = \frac{11}{18}$	1	
$P(X = 3) = {}^5C_3 \, q^2 p^3 = 10\left(\frac{11}{18}\right)^2\left(\frac{7}{18}\right)^3$		
$\underline{= 0.22 \ (2 \text{ s.f.})}$	2	Show your working; do not just state an answer.
For the binominal to be satisfactory the events must be independent and the value of p must be constant. These requirements would not be met if, for example, there was a change in circumstances such as severe weather conditions.	1 1	Think about how the conditions for a binomial distribution may not be satisfied.
2 (a) Let X be the number that arrive in an hour.		The clues to the distribution are the words 'at random' and 'rate of 5 per hour'.
Then $X \sim \text{Po}(5)$		
$P(X = 3) = e^{-5}\frac{5^3}{3!} = \underline{0.140 \ (3 \text{ s.f.})}$	2	
(b) Let Y be the number that arrive in 15 mins.		You need to change the variable to deal with the time interval of $\frac{1}{4}$ hour.
Then $Y \sim \text{Po}(1.25)$	1	
$P(\text{next patient arrives before 10.15 a.m.})$	1	
$= P(Y \geq 1)$		
$= 1 - P(Y = 0)$		
$= 1 - e^{-1.25}$		
$= \underline{0.713 \ (3 \text{ s.f.})}$	2	

Answer	Mark	Examiner's tip

3 Let $P(X=5)=p$ then $P(X=3)=2p$

Since $\Sigma P(X=x)=1$,

$\frac{1}{13}+\frac{3}{13}+2p+2p+2p+2p+p=1$

$9p=\frac{9}{13}$

$p=\frac{1}{13}$

x	1	2	3	4	5	6	7
$P(X=x)$	$\frac{1}{13}$	$\frac{3}{13}$	$\frac{2}{13}$	$\frac{2}{13}$	$\frac{1}{13}$	$\frac{2}{13}$	$\frac{2}{13}$

1 — State the theory that you are using.

It is easier to see the information in a table.

(a) Mean $=\Sigma x\, P(X=x)$

$=\frac{1}{13}(1+6+6+8+5+12+14)$

$=\underline{4}$ **2**

$E(X^2)=\Sigma x^2\, P(X=x)$

$=\frac{1}{13}(1+12+18+32+25+72+98)$

$=\frac{258}{13}$

$\mathrm{Var}(X)=E(X^2)-(\text{mean})^2$

$=3\frac{11}{13}$ **2**

(b) $E(X+3)=E(X)+3=7$

$\mathrm{Var}(X+3)=\mathrm{Var}(X)=3\frac{11}{13}$ **2**

This is a typical question relating to a discrete probability distribution, so make sure that you can apply the formulas accurately.

Remember that adding 3 to all the variables does not alter the variance, which is a measure of dispersion about the mean.

4 (a) Let W be the height of a foxglove growing in woodland, measured in cm.

Then $W \sim N(27.5, 3.5^2)$

$P(W>35)=P\left(Z>\dfrac{35-27.5}{3.5}\right)$ **2**

$=P(Z>2.143)$ **1**

$=\underline{0.016}$ (3 d.p.)   **1**

This question is testing your ability to standardise normal variables and use normal distribution tables.

Remember that $Z=\dfrac{X-\mu}{\sigma}$.

(b) Let R be the height, in cm, of a foxglove growing on the riverbank.

$R \sim N(32.0, 4^2)$

$P(R<35)=P\left(Z<\dfrac{35-32.0}{4}\right)$

$=P(Z<0.75)$ **1**

$=\underline{0.773}$ (3 d.p.)   **1**

It is helpful to be systematic about the way that you show your working.

Consider also whether diagrams would be helpful in checking the reasonableness of the Z value and the final probability.

Answer	Mark	Examiner's tip

(c) To find $P(W>R)$, i.e. $P(W - R > 0)$ consider
the distribution of $W - R$.

$$E(W - R) = E(W) - E(R) = 27.5 - 32.0$$

$$= -4.5 \qquad \qquad 1$$

$$\text{Var}(W - R) = \text{Var}(W) + \text{Var}(R)$$

$$= 3.5^2 + 4^2 = 28.25 \qquad 1$$

Since W and R are both normally
distributed, and independent

$$W - R \sim \text{N}(-4.5,\ 28.25) \qquad 1$$

$$P(W - R > 0) = P\left(Z > \frac{0 - (-4.5)}{\sqrt{28.25}}\right) \qquad 1$$

$$= P(Z > 0.847) \qquad 1$$

$$= 0.199 \ (3 \text{ d.p.}) \qquad 1$$

A common mistake is to subtract
the variances. In this instance you
should realise that you have made
an error, because Var(*W–R*) would
be negative, which is impossible.

5 (a) Poisson (λ) where λ is the mean number
 of sales per week.   **1**

Recognise the distribution of
random events.

(b) (i) If X is the number sold in the first
showroom, and Y the number in the
second,

$$X \sim \text{Po}(2.4),\ Y \sim \text{Po}(3.6)$$

$$P(X = 3) = \frac{2.4^3}{3!}\ e^{-2.4} = 0.2090\ldots \qquad 1$$

$$P(Y = 5) = \frac{3.6^5}{5!}\ e^{-3.6} = 0.1376\ldots$$

If X and Y are independent, required   **2**
probability = $P(X = 3) \times P(Y = 5)$

$$= 0.0288 \ (3 \text{ s.f.}) \qquad 1$$

There is no need to approximate
these values.

You probably multiplied the
probabilities, but did you realise
the significance of the word
'independent'?

(ii) Let T be total sold in two showrooms,

then $T \sim \text{Po}(2.4 + 3.6)$, i.e. $T \sim \text{Po}(6)$ **1**

$$P(T < 8)$$

$$= e^{-6}\left(1 + 6 + \frac{6^2}{2!} + \frac{6^3}{3!} + \frac{6^4}{4!} + \frac{6^5}{5!} + \frac{6^6}{6!} + \frac{6^7}{7!}\right)$$

$$= 0.744 \ (3 \text{ s.f.}) \qquad 1$$

It is important to remember that the
sum of two independent Poisson
variables is also Poisson.

Use cumulative probability tables if
they are available.

Answer	Mark	Examiner's tip

(iii) Let W be the amount sold by the first showroom in 10 weeks, then
$W \sim Po(24)$

We use a normal approximation, since λ is large, so $W \sim N(24,24)$ — **1** — An approximation has been hinted at in the question.

Assume that sales in each week are independent of each other.

P(at least 30)
$$=P(W \geq 29.5) = P\left(Z \geq \frac{29.5 - 24}{\sqrt{24}}\right)$$ — **1**

Remember to use the continuity correction. The words 'at least' are very important here.

$$= P(Z \geq 1.123)$$ — **1**

$$= 0.131 \ (3 \text{ s.f.})$$ — **1**

6 $f(t) = kt, \ 0 < t \leq 4$

(a) (i) $1 = \int_0^4 kt \, dt = \frac{k}{2}\left[t^2\right]_0^4 = 8k$

$$\Rightarrow k = \tfrac{1}{8} = 0.125$$ — **1**

Remember that the total area under the curve is 1. Instead of integrating, you could draw a sketch and find the area of the triangle so formed.

(ii) $E(T) = \int t f(t) dt = \tfrac{1}{8}\int_0^4 t^2 \, dt$

$$= \tfrac{1}{8}\left[\frac{t^3}{3}\right]_0^4 = \tfrac{8}{3} \text{ minutes}$$ — **2**

It is useful to state the formulas being used and show all your working.

$$E(T^2) = \int t^2 f(t) dt = \tfrac{1}{8}\int_0^4 t^3 \, dt$$

$$= \tfrac{1}{8}\left[\frac{t^4}{4}\right]_0^4 = 8$$ — **1**

$$\text{Var}(T) = E(T^2) - \left(\tfrac{8}{3}\right)^2$$

$$= \tfrac{8}{9}$$ — **1**

Standard deviation

$$= \sqrt{\tfrac{8}{9}} = \frac{2\sqrt{2}}{3} \text{ minutes}$$ — **1**

Remember to give the units in your answer.

(iii) $P(3 < T < 4) = \int_3^4 \tfrac{1}{8} t \, dt = \tfrac{1}{16}\left[t^2\right]_3^4 = \tfrac{7}{16}$ — **1**

(iv) If X is the number out of the 5, lasting between 3 and 4 minutes, then

$$X \sim B\left(5, \tfrac{7}{16}\right)$$

This is a binomial distribution, since there are 5 independent events with constant probability of $\tfrac{7}{16}$ of 'success'.

$$P(X = 3) = {}^5C_3\left(\tfrac{9}{16}\right)^2\left(\tfrac{7}{16}\right)^3$$
$$= 0.265 \ (3 \text{ s.f.})$$ — **3**

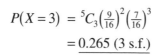

Answer	Mark	Examiner's tip

(b) (i) $P(0 < T < 1) = \frac{1}{16}\left[t^2\right]_0^1 = \frac{1}{16}$

$P(1 < T < 2) = \frac{1}{16}\left[t^2\right]_1^2 = \frac{3}{16}$

$P(2 < T < 3) = \frac{1}{16}\left[t^2\right]_2^3 = \frac{5}{16}$

Length of call (mins)	Probability	Cost, c (pence)
0–1	$\frac{1}{16}$	6
1–2	$\frac{3}{16}$	10
2–3	$\frac{5}{16}$	14
3–4	$\frac{7}{16}$	18

3

For (b) you need to use the distribution given in (a). Use the result obtained in (iii), with the appropriate limits, or set up the cumulative function $F(x) = \dfrac{x^2}{16}$.

Check that the sum of the probabilities is 1.
Assume that 0–1 does not include 1 complete minute.

(ii) Let C be the cost of the call, in pence.

$E(C) = \Sigma c P(C = c)$

$\qquad = \frac{1}{16}(6 + 30 + 70 + 126) = \underline{14.5 \text{ pence}}$ **1**

$E(C^2) = \Sigma c^2 P(C = c)$

$\qquad = \frac{1}{16}(36 + 300 + 980 + 2268) = 224$ **1**

$\text{Var}(C) = E(C^2) - E^2(C) = 224 - 14.5^2$

$\qquad = 13.75$ **1**

$\underline{\text{Standard deviation} = 3.71 \text{ pence (2 d.p.)}}$ **1**

The question now concentrates on a discrete probability distribution – be prepared for a change of emphasis.

7 $P(A) = 0.37,\ P(B) = 0.004$

(a) If X is the number that are type A, then X follows a binomial distribution with $n = 8$, $p = 0.37,\ q = 1 - p = 0.63$

$P(X \geq 2) = 1 - P(X < 2)$ **1**

$\qquad = 1 - \big(P(X = 0) + P(X = 1)\big)$

$\qquad = 1 - \big((0.63)^8 + 8(0.63)^7(0.37)\big)$

$\qquad = \underline{0.859 \text{ (3 s.f.)}}$ **2**

You are not told which distribution to use, so you need to recognise that you have independent events with p constant.

(b) $P(A \text{ or } B) = P(A) + P(B)$ since A and B are mutually exclusive. **1**

So $P(A \text{ or } B) = 0.374$

If Y is the number of type A or B in 200,

You need to apply some probability theory, before considering the distribution.

Answer	Mark	Examiner's tip

then Y follows a binomial distribution with $n = 200$, $p = 0.374$.

Since n is large and p is close to 0.5, we can use a normal approximation with mean $np = 200(0.374) = 74.8$ and variance $npq = (74.8)(1–0.374) = 46.8248$.

With the continuity correction, consider

$$P(X > 80.5) = P\left(Z > \frac{80.5 - 74.8}{\sqrt{46.8248}} \right)$$

$$= P(Z > 0.8329 \ldots)$$

$$= \underline{0.203} \text{ (3 s.f.)}$$

2
2
State the actual distribution before considering the approximation.

Justify your use of an approximation.

2
1
1
The continuity correction is needed because a continuous distribution is being used as an approximation to a discrete one.

(c) Let W be the number of type B in 300. Then W follows a binomial distribution with $n = 300$, $p = 0.004$, so $np = 1.2$

1
Again, remember to state the actual distribution before considering the approximation.

Now since n is large and p is very small, such that $np \approx npq$, we can use a Poisson approximation where $W \sim \text{Po}(1.2)$

This is an important feature of this approximation.

$$P(W \geq 4) = 1 - P(W < 4)$$

1

$$= 1 - \left(e^{-1.2} + 1.2e^{-1.2} + \frac{1.2^2}{2!} e^{-1.2} + \frac{1.2^3}{3!} e^{-1.2} \right)$$

2
Use cumulative probability tables if they are available.

$$= \underline{0.0338} \text{ (3 s.f.)}$$

1

8 X is the number of wet days in 10 and $X \sim \text{B}(10, p)$

$H_0 : p = \frac{2}{7}$

$H_1 : p > \frac{2}{7}$ (and wet days are more likely to be a Saturday or Sunday)

The *first* step in carrying out a significance test is to state the hypotheses. The distribution, according to H_0, can then be defined.

According to H_0, $X \sim \text{B}\left(10, \frac{2}{7}\right)$

The sample value is $x = 5$, so reject H_0 if $P(X \geq 5) < 0.1$.

You need to find out whether the sample value lies in the upper tail 10% of the distribution.

$P(X \geq 5) = 1 - P(X \leq 4)$

$$= 1 - \left(\left(\tfrac{5}{7}\right)^{10} + 10 \left(\tfrac{5}{7}\right)^9 \left(\tfrac{2}{7}\right) + 45\left(\tfrac{5}{7}\right)^8 \left(\tfrac{2}{7}\right)^2 \right.$$

2

$$\left. + 120\left(\tfrac{5}{7}\right)^7 \left(\tfrac{2}{7}\right)^3 + 210\left(\tfrac{5}{7}\right)^6 \left(\tfrac{2}{7}\right)^4 \right)$$

Use cumulative probability tables if they are available.

$$= \underline{0.127} \text{ (3 s.f.)}$$

2

Since $0.127 > 0.1$, there is not enough evidence to reject H_0 and the gardener's claim is not upheld.

2
When you make your conclusion, relate it to the given situation.

Answer	Mark	Examiner's tip

9 (i) Let X be the length, in cm, of a key.

$$\hat{\mu} = \bar{x} = \frac{\Sigma x}{n} = \frac{250.5}{50} = \underline{5.01}$$

1

Show your working, stating the formulas. Remember that an unbiased estimate of the population mean is the sample mean.

$$\hat{\sigma}^2 = \frac{n}{n-1}\left(\frac{\Sigma x^2}{n} - \bar{x}^2\right)$$

$$= \frac{50}{49}\left(\frac{1255.0290}{50} - (5.01)^2\right)$$

Check the formula for the unbiased estimate of the variance in your reference booklet as there are several formats.

$$= \underline{0.000\ 489\ 8}\ (4\ \text{s.f.})$$

2

(ii) 95% confidence interval for μ is

$$\bar{x} \pm 1.96 \frac{\hat{\sigma}}{\sqrt{n}} = 5.01 \pm 1.96 \frac{\sqrt{0.000\ 489\ 8}}{\sqrt{50}}$$

2

Learn the formula.

$$= (5.004,\ 5.016)\ (3\ \text{d.p.})$$

2

(iii) $H_0 : \mu = 5.00$

1

$H_1 : \mu > 5.00$ (the keys are too long)

1

To test the assertion that the keys are too long, a one-tailed test is needed.

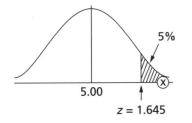

According to H_0, and applying the Central Limit Theorem,

The Central Limit Theorem is needed, since we do not know the distribution of X.

$$\bar{X} \sim N\left(\mu, \frac{\hat{\sigma}^2}{n}\right)$$

i.e. $\bar{X} \sim N(5.00,\ 0.000\ 009\ 7)$

Reject H_0 if $Z > 1.645$

Alternatively, investigate whether $P(X > 5.01) < 0.05$; if so, reject H_0.

where $Z = \dfrac{5.01 - 5.00}{\sqrt{0.000\ 009\ 7}}$

$$= 3.20$$

2

$\underline{3.20 > 1.645}$, so we reject H_0 and

conclude that the keys are too long.

2

Answer	Mark	Examiner's tip

10(a) Let X be the number of girls in a family of 4.

H_0 : X is distributed binomially with $n = 4$, $p = 0.5$

H_1 : X does not follow this distribution.   **1**

It is important that you state the null hypothesis clearly. From it, the expected frequencies are calculated.

The expected frequencies in 200 families, according to H_0, are calculated as follows:

$P(0) = (0.5)^4 = 0.0625 \Rightarrow E_0 = 12.5$

$P(1) = 4(0.5)^4 = 0.25 \Rightarrow E_1 = 50$

$P(2) = 6(0.5)^4 = 0.375 \Rightarrow E_2 = 75$

To find the expected frequencies multiply the probabilities by 200.

Since the distribution is symmetric,

$E_3 = 50$, $E_4 = 12.5$ **3**

It is useful to recognise the symmetry.

O	15	68	69	38	10
E	12.5	50	75	50	12.5

All the expected frequencies are greater than 5, so there is no need to pool any classes.

Number of degrees of freedom = $5 - 1 = 4$.

The critical value of χ^2 (with 4 degrees of freedom) at the 5% level is 9.49, so we will reject H_0 if the χ^2 value is greater than 9.49.   **2**

There are 5 classes, and 1 restriction (that totals agree).

$$\chi^2 = \sum \frac{(O-E)^2}{E}$$

$$= \frac{2.5^2}{12.5} + \frac{18^2}{50} + \frac{6^2}{75} + \frac{12^2}{50} + \frac{2.5^2}{12.5}$$

$$= 10.84$$ **2**

Show your working.

Since $10.84 > 9.49$, reject H_0 and conclude that the binominal distribution with $n = 4$, $p = 0.5$ does not fit.   **2**

Relate the conclusion to the practical situation.

(b) For the data in the given table

$$\text{mean} = \frac{\Sigma fx}{\Sigma f} = \frac{360}{200} = 1.8$$

$$\Rightarrow np = 1.8$$

$$p = 1.8 \div 4 = \underline{0.45}$$ **2**

Answer	Mark	Examiner's tip

(c) There are now 2 restrictions (totals agree and *p* is estimated from the sample)

Number of degrees of freedom = 5 − 2 = 3, and the critical value of χ^2 is 7.81.

2 Note that the number of degrees of freedom has changed.

Since 2.47 < 7.81, we do not reject the null hypothesis (that *X* follows a binomial distribution with *n* = 4, *p* = 0.45).

So this binomial model is a good fit. **2**

(d) The number of girls in a family follows a binomial distribution, and 0.45 is a good estimate of the probability of having a girl. **2**

11 (i)

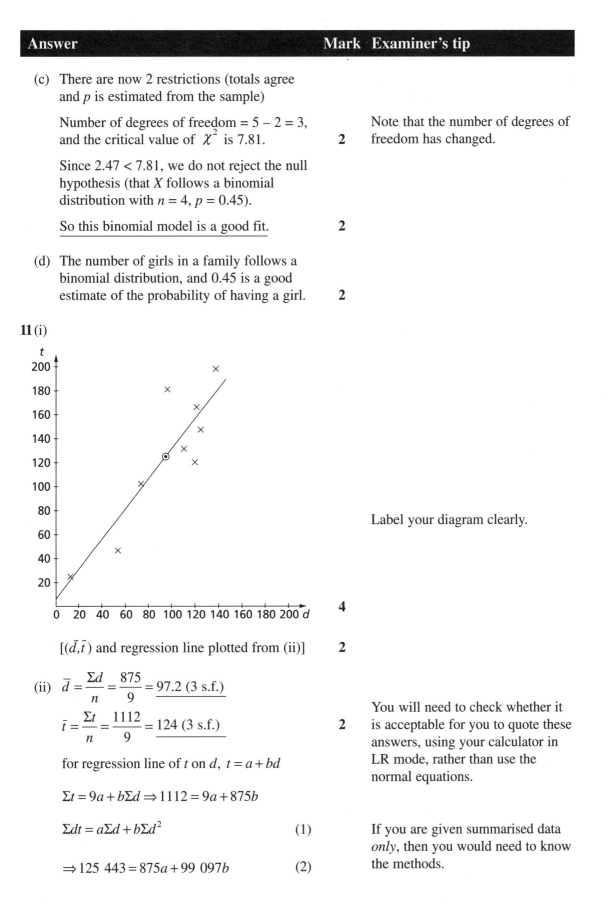

4

$[(\bar{d}, \bar{t})$ and regression line plotted from (ii)] **2**

Label your diagram clearly.

(ii) $\bar{d} = \dfrac{\Sigma d}{n} = \dfrac{875}{9} = 97.2$ (3 s.f.)

$\bar{t} = \dfrac{\Sigma t}{n} = \dfrac{1112}{9} = 124$ (3 s.f.) **2**

You will need to check whether it is acceptable for you to quote these answers, using your calculator in LR mode, rather than use the normal equations.

for regression line of *t* on *d*, $t = a + bd$

$\Sigma t = 9a + b\Sigma d \Rightarrow 1112 = 9a + 875b$

$\Sigma dt = a\Sigma d + b\Sigma d^2$ (1)

$\Rightarrow 125\ 443 = 875a + 99\ 097b$ (2)

If you are given summarised data *only*, then you would need to know the methods.

Answer	Mark	Examiner's tip

solving these equations gives

$a = 3.43$ (3 s.f.)

$b = 1.24$ (3 s.f.)

so $\underline{t = 3.43 + 1.24d}$ **2**

(iii) When $d = 100$, $t = 3.43 + 124 \approx \underline{127 \text{ min}}$ **2**

(iv) $r = \dfrac{s_{dt}}{s_d s_t}$ where $s_{dt} = \dfrac{\Sigma dt}{9} - \bar{d}\bar{t}$

$s_d^{\,2} = \dfrac{\Sigma d^2}{9} - \bar{d}^2$

r can be obtained directly from the calculator, so use it to check.

$s_t^{\,2} = \dfrac{\Sigma t^2}{9} - \bar{t}^2$

$r = \dfrac{1925.7654}{(39.4...)(54.5...)} = \underline{0.894}$ (3 d.p.)   **2**

Also check the format of the formulas you will be given in the examination.

This value is close to 1 and shows a good positive linear correlation between *d* and *t*, inferring that *t* is increasing with *d*.   **2**